城市写意
——建筑师的创意

[英]史蒂夫·鲍克特 著

马 嵩 张云峰 译

中国建筑工业出版社
CHINA ARCHITECTURE & BUILDING PRESS

著作权合同登记图字：01-2017-4366号

图书在版编目（CIP）数据

城市写意：建筑师的创意／（英）鲍克特著；马嵩等译．—北京：中国建筑工业出版社，2017.12

ISBN 978-7-112-21303-0

Ⅰ.①城… Ⅱ.①鲍… ②马… Ⅲ.①建筑画－作品集－英国－现代 Ⅳ.①TU204

中国版本图书馆CIP数据核字（2017）第243509号

©2017 Steve Bowkett.
Steve Bowkett has asserted his right under the Copyright, Design and Patent Act 1988 to be identified as the Author of this Work.
Translation ©2017 China Architecture and Building Press
This book was designed, produced and published in 2017 by Laurence King Publishing Ltd under the title *Archidoodle City*: *An Architect's Book*.
This Translation is published by arrangement with Laurence King Publishing Ltd. for sale/distribution in The Mainland (part) of the People's Republic of China (excluding the territories of Hong Kong SAR, Macau SAR and Taiwan Province)

本书由英国Laurence King 出版社授权翻译出版

责任编辑：程素荣
责任校对：王　烨

城市写意——建筑师的创意
[英] 史蒂夫·鲍克特　著
马　嵩　张云峰　译
＊
中国建筑工业出版社出版、发行（北京海淀三里河路9号）
各地新华书店、建筑书店经销
北京锋尚制版有限公司制版
北京中科印刷有限公司印刷
＊
开本：787×1092毫米　1/12　印张：13⅓　字数：243千字
2018年2月第一版　　2018年2月第一次印刷
定价：48.00元
ISBN 978 – 7 – 112 – 21303 – 0
　　　（31016）

版权所有　翻印必究
如有印装质量问题，可寄本社退换
（邮政编码100037）

ARCHI-DOODLE CITY

AN ARCHITECT'S ACTIVITY BOOK

本书适用于不同年龄段的建筑师

STEVE BOWKETT

也门的希巴姆城（Shibam）被称为"沙漠中的曼哈顿"，它拥有大约500座由泥土砖构成的塔式建筑群，这些塔楼的高度从5~11层不等。从16世纪开始，这些塔楼经历了多次修复。希巴姆城市是紧凑式城市规划第一个和最优秀的范例。

Introduction（导言）

目前世界人口的一半以上居住在城市中心，这个数字还在逐年增长。如何应对城市人口爆炸性增长以及重塑城市生活是建筑师与城市规划师最关注的问题。《城市写意》这本书既回顾了城市发展的历程，又能够启发对城市未来的畅想。像我的第一本书《建筑写意》一样，这是一本能够吸引你对各类建筑进行设计、速写、涂色和涂鸦的互动书籍，只不过这次的焦点是城市。我使用大量的城市及其细节作为每一个挑战的起点：涉及范围从古代到近代，从小城市到大城市，其中的挑战包含了学术和娱乐，当然最重要的是能够让你在绘制过程中获得快乐。

你可以随意地使用任何风格或你喜欢的任何媒介进行绘画。我绘制的所有图画都尽量保持干净整洁并力图清晰，以便于你在提供的空白页面内用不同的工具去尝试。所有的图都是黑白色的，你也可以随意将其作为一本着色书使用。许多学生将《建筑写意》视作一本重要的组合初级读本，我希望《城市写意》也能够在激励所有读者去绘制和梦想未来城市环境方面发挥同样重要的作用。

关于作者

史蒂夫·鲍克特是一位对优秀设计充满激情的作者。他在众多所大学和学院从事教学和建筑实践超过25年，目前是英国伦敦南岸大学（London South Bank University）的高级讲师。他曾在英国皇家艺术学院（the Royal College of Art in London）和伦敦中央理工学院（the Polytechnic of Central London）学习建筑学。史蒂夫的上一本著作《建筑写意》已经出版了10种语言。

史蒂夫目前与他的妻子简和女儿们——佐伊、萨迪和菲比住在白金汉郡，一起生活的还有一条狗、两只猫和一百条鱼。尽管仍在追求一种不期而遇的生活，史蒂夫偶尔也会找时间无所事事。

Equipment（设备）

这些是你在本书中可以考虑使用的基本工具

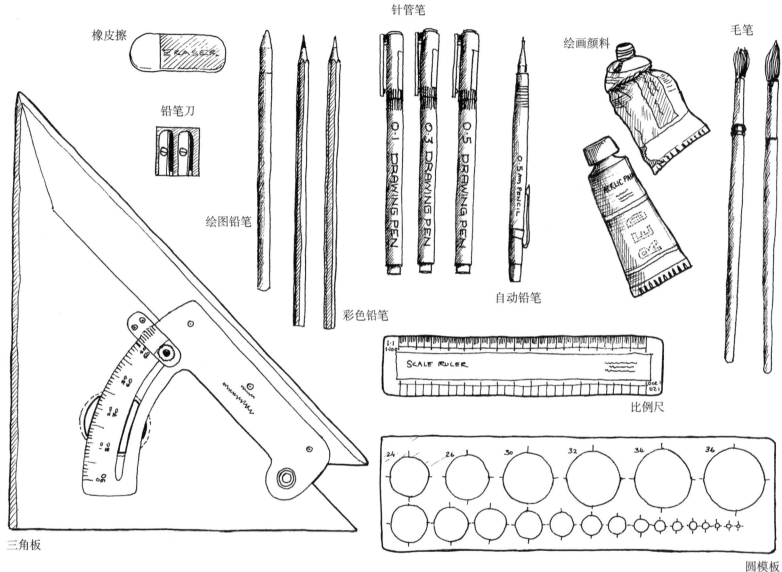

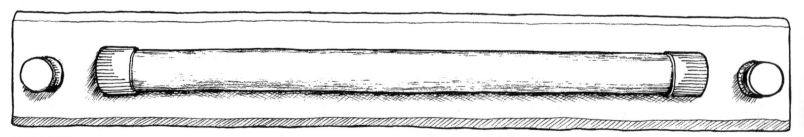

Techniques（技法）

本页展示了我在本书绘图中所使用的一些技术选择，这些简单的技法有助于你构建结构和形式，添加阴影和增加密度，并创建一系列不同的材质效果。

平行线　　　　　　　　　　　　　　　　　　　　　交叉线

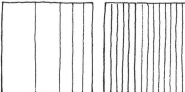 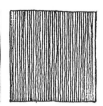  

点画技法

   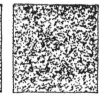

材质效果：珊瑚；沉积物；草地；大理石；叶子；植被；有波纹的水面；表面纹理；静止的水面

 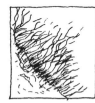 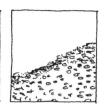    

材质效果：水；水泥渲染（粉刷）；瓦片屋顶；植栽；铺装；砖石；织物；岩石

 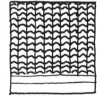  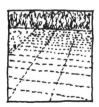   

Sketch perspectives（草图透视）

以下是透视画法的入门指南，而并非全面的画法指导。一点透视是展示内部空间的常用方法——下图是一个街道内景的例子。在一点透视图中，所有绘图要素都向水平面上唯一的一点汇聚（灭点 - Vanishing Point），该点与观察者视线中心重合。随着画中的要素越来越远，它们看起来愈加集中且微小。

一点透视

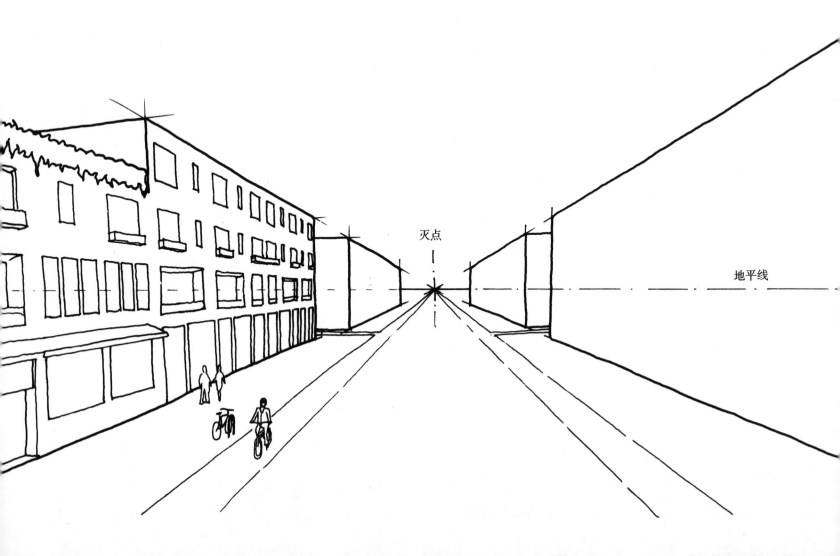

两点透视是展示建筑三维体块的重要方法。下面的例子显示了地平线（或者视线）移动过程中（例如向上看或向下看），它是如何改变观察者和建筑之间的关系的。你可以发现，地平线上有两个同时存在的灭点，垂直的建筑线永远与地平线保持直角且彼此平行。拉近两个灭点彼此的距离将会使建筑变形，并产生倾斜的视觉感受。反之，分离两个灭点将会使整个图像变平。

两点透视

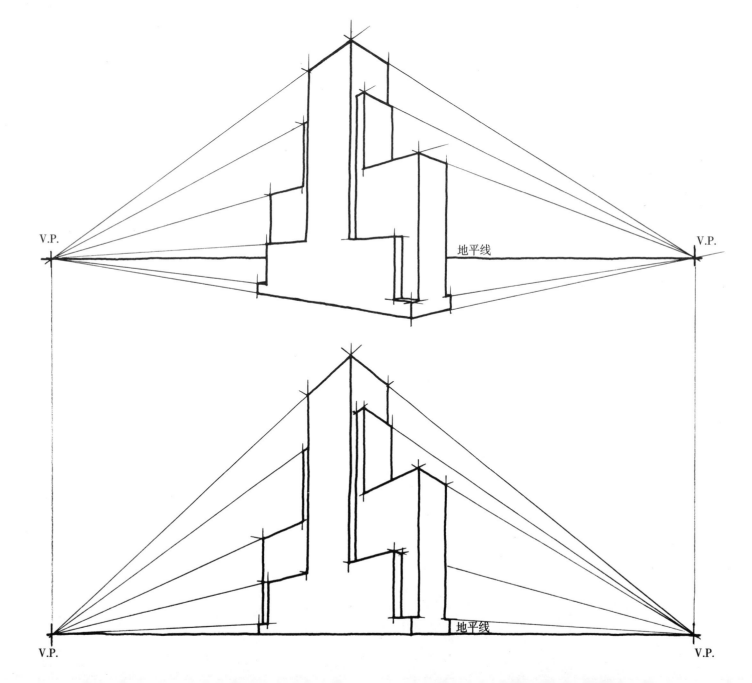

This waterfront panorama of the **city of Shanghai** in China is in a constant state of flux, with new buildings changing the skyline every year.

中国城市上海的滨水界面景观一直处于变化状态，因为每年都有新增建筑改变它的天际线。

你将如何利用提供的空白空间改变这个城市的特征？

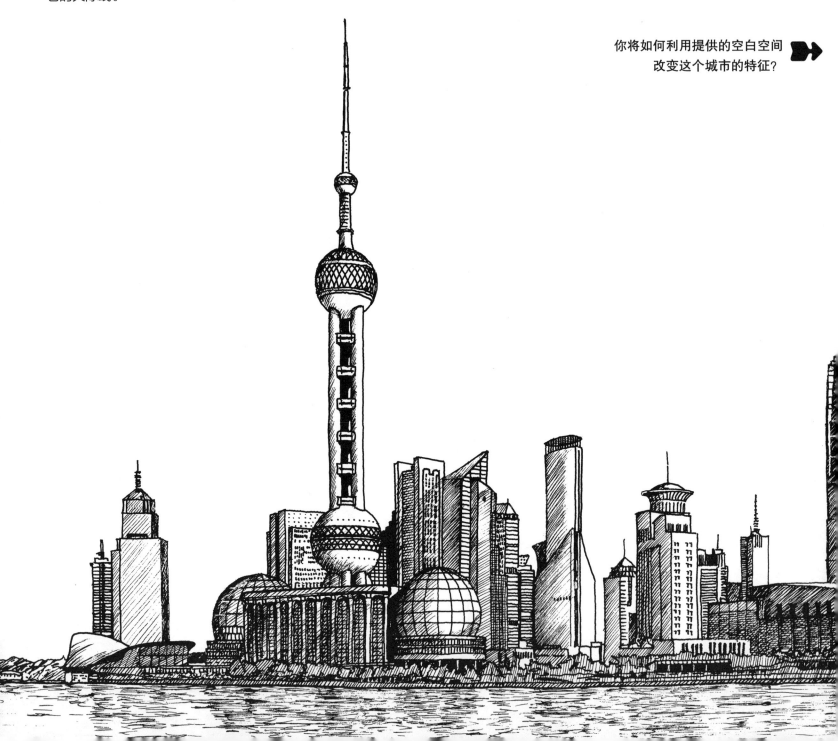

Shopping and **shop fronts** form a major part of our experience of cities. Here are some examples of shop fronts that either celebrate or disregard the goods or services being sold.

购物和店面是我们感受城市的重要组成部分。这里有一些店面的例子，它们要么展示、要么无视正在销售中的商品或服务。

在提供的空间内，为某特定商品设计一个店面。

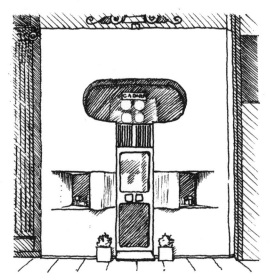

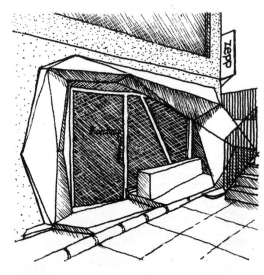

左图：沙龙·米特迈尔理发店，林茨，奥地利。X建筑师事务所，2008年。

上图：雷特尔蜡烛商店，维也纳，奥地利。霍莱恩，1965年。

下图：康赛普商店，服装零售，香港。徐君俊，2010年。

Derinkuyu Underground City is an ancient

(eighth to seventh century BCE) underground city of the Median Empire in the Derinkuyu district of Turkey's Nevsehir province. Extending down over many storeys to a depth of around 60 metres (200 feet), it was large enough to shelter at least 20,000 people, together with their livestock and food supplies. This is the largest excavated underground city in Turkey and is just one of several underground complexes found across the Cappadocia region.

德林克尤地下城是米提亚帝国时期的一处古代地下城，它位于土耳其内夫谢希尔省的德林库尤地区。由于向地下延伸数层且深度达约60米，这座地下城的空间足以庇护2万人，包括这些人的牲畜和食物供给。它是土耳其目前挖掘出来的最大地下城，也是卡帕多西亚地区挖掘出的众多地下设施之一。

完成你自己的地下城剖面绘制

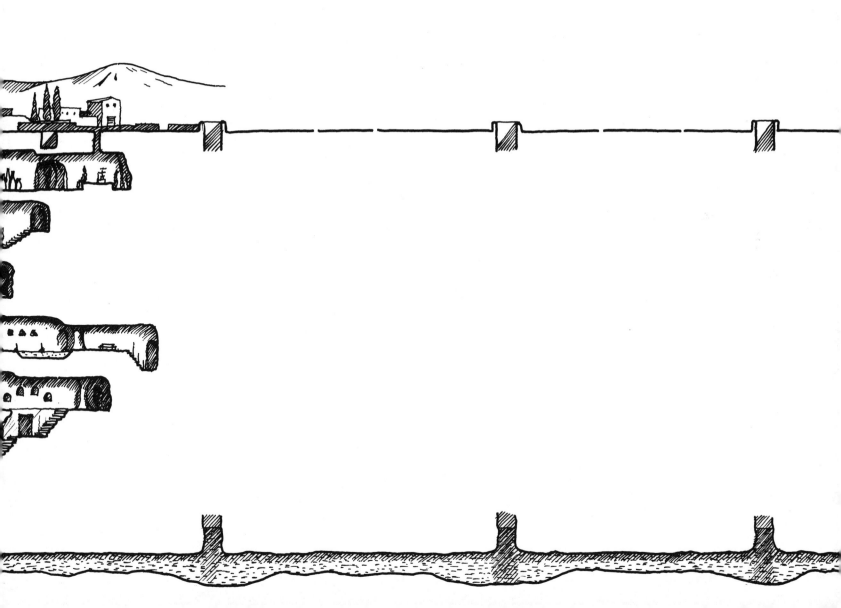

The question of **urban context** within cities has always been a much-debated issue. This example of a remodelled brownstone, designed by architect Edward Durrell Stone for his own use in 1956 in New York City, has been the subject of much controversy due to its stark contrast with its neighbours. It is currently protected as an official city landmark.

城市文脉一直以来都是一个颇具争议的话题。这个改建高级公寓的案例，由建筑师爱德华·达雷尔·斯通于1956年在纽约为自己使用而设计。该设计因其与邻里之间显著的对比而引起较大争议，目前它作为官方城市地标而受到保护。

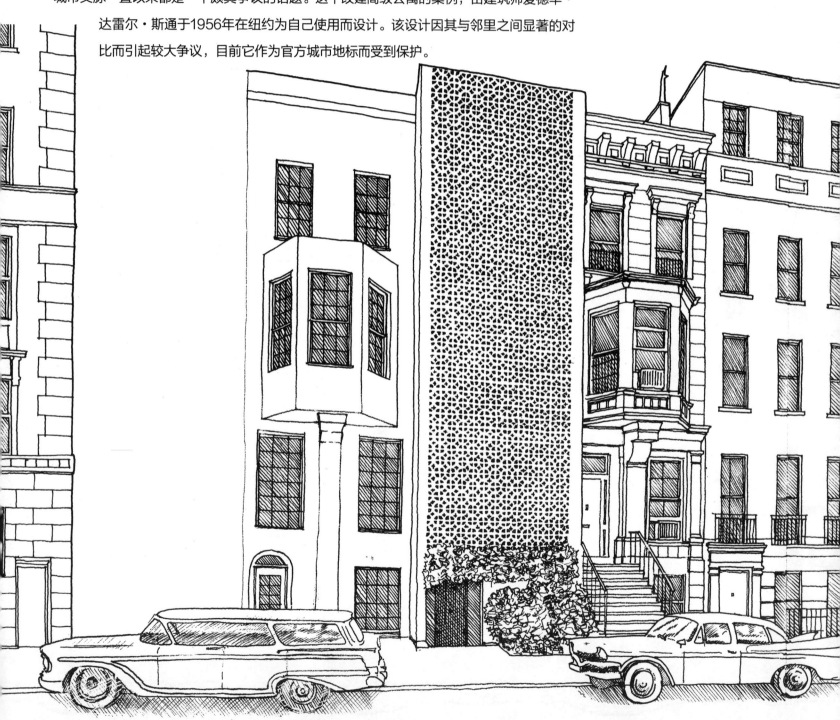

利用阳台之间的间隙,你将设计什么样的立面?传承城市文脉的立面?或者不是?

These silhouettes depict **famous cities.** Can you name them, and devise your own city images?

这些剪影描绘了一些著名的城市。你能认出它们吗?

能否设计出你自己的城市形象?

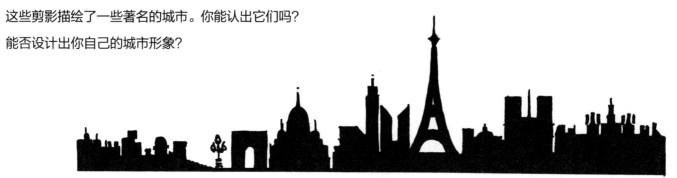

勾勒出你自己的城市剪影，并且努力创造出一些令人难忘的标志性建筑。

The **Hoover Dam Bypass Bridge** (officially the Mike O'Callaghan–Pat Tillman Memorial Bridge) is an arch bridge that spans 323 metres (1060 feet) across the Colorado River between Arizona and Nevada. The bridge was constructed to re-route US Route 93 away from the top of the Hoover Dam, and it was completed in 2010.

胡佛水坝大桥（官方名为麦克·奥卡拉汉-帕特·提尔曼纪念大桥）是一座跨度达323米的拱桥，它横跨美国亚利桑那州和内华达州之间的科罗拉多河。该桥因胡佛大坝上的美国93号公路改道而修建，并于2010年完成。

如果你有机会设计这么一个重要的构筑物，你会设计一个什么样的桥梁？

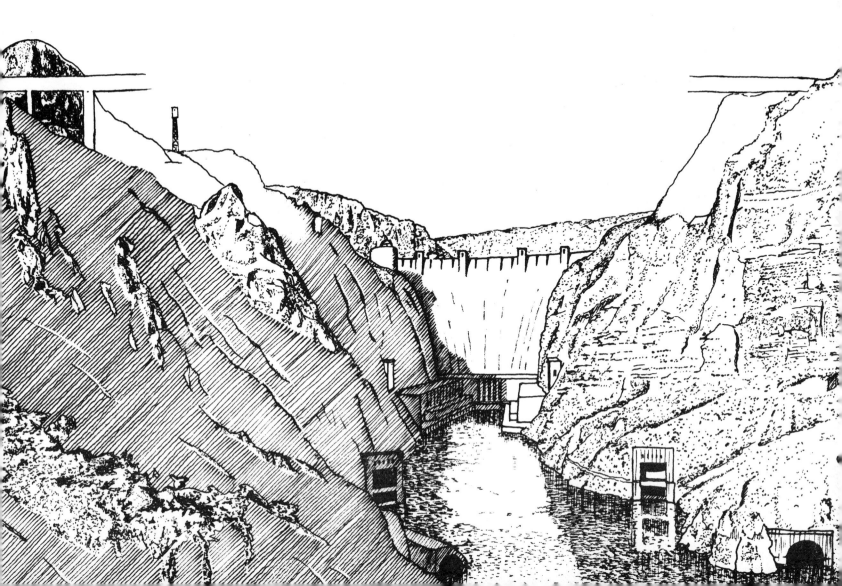

My daughter Phoebe and I decided to make a **model city** using utensils, containers and condiments from the kitchen. This partially completed drawing is one half of what we did.

我的女儿菲比和我决定用厨房里的餐具、容器和调味品制作一个模型城市，下边的部分完成的图画只是我们所做工作一半。

尝试重复这些练习并且在我的图中加入你自己的版本。

Above the rooftops of Manhattan is a landscape dominated by an estimated 17,000 **water towers.** With the majority still in use (a few have been converted into roof extensions) an educational group came up with the idea of using them as a canvas to promote the global water crisis. In 2014 'the Water Tank Project' was launched and invited over 100 acclaimed artists and New York City public school students to propose a series of wraparound artworks to cover these structures as part of this awareness campaign.

曼哈顿的屋顶景观由大约17000个水塔主导。由于这些水塔中的大多数仍在使用中（一些已转换成屋顶的延伸），一个教育小组想出一个利用水塔作为画布的主意，并将其应用于改善全球水危机活动。2014年，一项名为"水箱项目"的宣传运动开始启动，该项目邀请了超过100名著名艺术家和纽约市公立学校的学生来设计一系列覆盖这些屋顶结构的全景艺术作品。

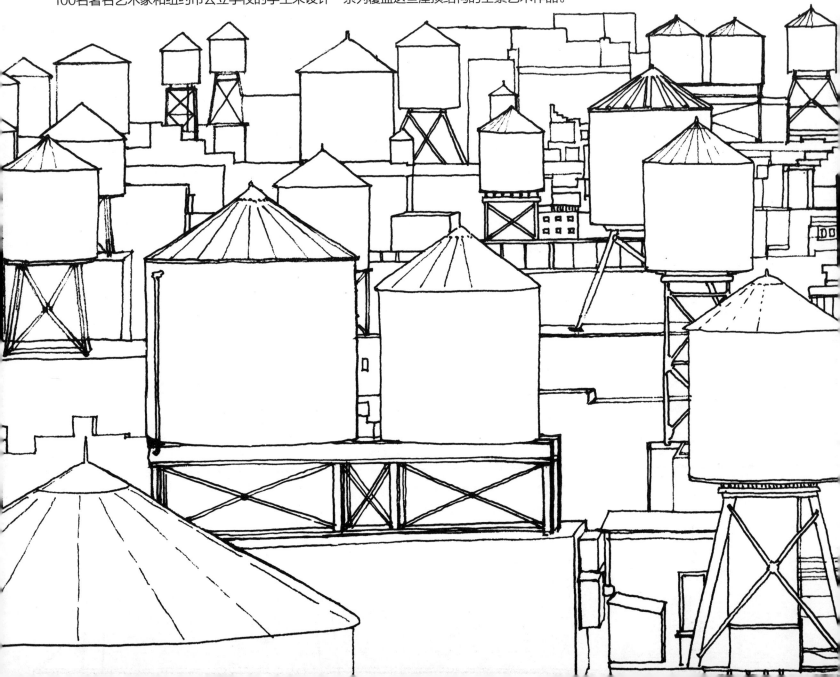

针对这个议题或者其他公共议题,你对这些水塔上的空白画布有什么建议?

In recent years the simple **public bench** has literally taken a twist. These three examples elevate this humble and pragmatic piece of street furniture to the level of sculpture.

近年来，简单的公共长椅已经发生了根本性的转变。以下三个案例见证了这些谦卑务实的街道家具向艺术品的提升。

"通心粉"长椅，帕布鲁·雷诺索，2006年。

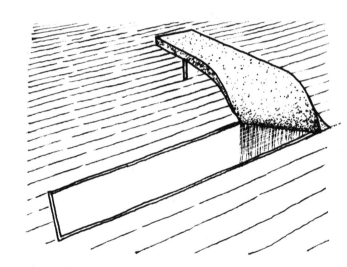

"蓝色地毯"长椅，赫斯维克事务所，2002年。

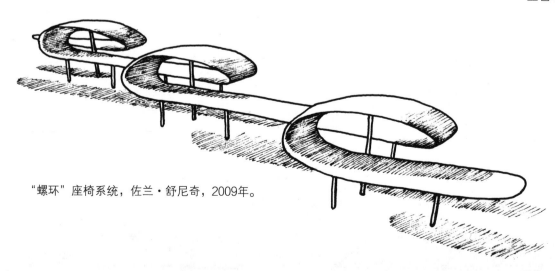

"螺环"座椅系统，佐兰·舒尼奇，2009年。

如果你受邀设计一个雕塑式公共长椅,它将是什么样子的?你要考虑材料和日常磨损。

This famous architects' costume party, **the Beaux-Arts Ball,** held in New York in 1931, invited the designers to dress up as their own buildings.

1931年，在纽约举办的著名建筑师的化装舞会——学院派舞会，邀请设计师装扮成自己的建筑物。

在提供的模板上设计并绘制你希望的建筑。 ➡

伊利雅克·卡恩装扮成斯夸波大厦，威廉·范·阿伦装扮成克莱斯勒大厦，拉夫·沃克装扮成华尔街1号。

对页图：爱奥尼柱式服装来自大卫·鲍伊1986年的电影"真实的故事"。

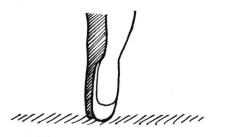

Positioned in one corner of Trafalgar Square, London, **the Fourth Plinth** remained devoid of a sculpture for 150 years after its creation. In 1999 the Royal Society of Arts decided, rather than commission a permanent sculpture, they would use the plinth to display temporary artworks by a succession of artists.

位于伦敦特拉法加广场的一个角的第四基座，从建成之日起，仍然缺失雕塑的状态已经150年了。1999年，皇家艺术学会与其委托制作一个永久性雕塑，还不如决定使用这个基座来展示一系列艺术家的临时艺术品。

你将为第四基座带来什么设计？

"纪念碑"。雷切尔·怀特瑞德，2001年。

"哈恩/公鸡"。凯塔琳娜·弗里奇，2013年。

The 180 **towers of Medieval Bologna,** built in Italy between the twelfth and thirteenth centuries, are believed to have been constructed by wealthy feuding Bolognese families. The absence of state control gave way to a lawless society in which the families needed to build taller and taller towers to defend themselves against each other. Alas, only two towers now remain. Imagine what the city would be like now if the towers had survived. Maybe new friendships would have formed?

12~13世纪建于意大利的180个中世纪博洛尼亚塔楼,据称是由富裕但处于持续争斗中的多个博洛尼亚家族建造而成。国家管控的缺失造成了当时缺乏法制的社会,生活在其中的家族们需要建设越来越高的塔楼来保护自己不受他人的攻击。可惜的是,目前只剩下两座塔楼了。想象一下,如果所有的塔楼都幸存下来,那么这个城市将会是怎样的。或许会形成新的友好关系?

你将如何重新构建这些城市家族之间的联系,也许通过不同类型桥梁横跨塔楼的方式?

The **Nakagin Capsule Tower,** designed by Kisho Kurokawa, was completed in Tokyo, Japan, in 1972. This building is one of the few remaining examples of Japanese Metabolism and was the first permanent example of capsule architecture. The living capsules measure 2.3 x 3.8 x 2.1 metres (7 1/2 x 12 1/2 x 7 feet) and are attached to two concrete service towers.

位于日本东京的中银胶囊大厦由黑川纪章设计，于1972年建成。这栋建筑是日本现存的少数"新陈代谢派"代表作之一，并且是第一个永久性胶囊建筑案例。这些居住胶囊的尺寸为2.3米×3.8米×2.1米，并被连接在两个混凝土主体塔楼上。

你将在这两栋垂直楼梯和电梯塔上连接什么样的居住单元？

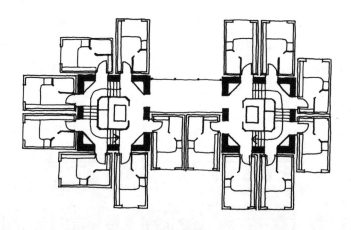

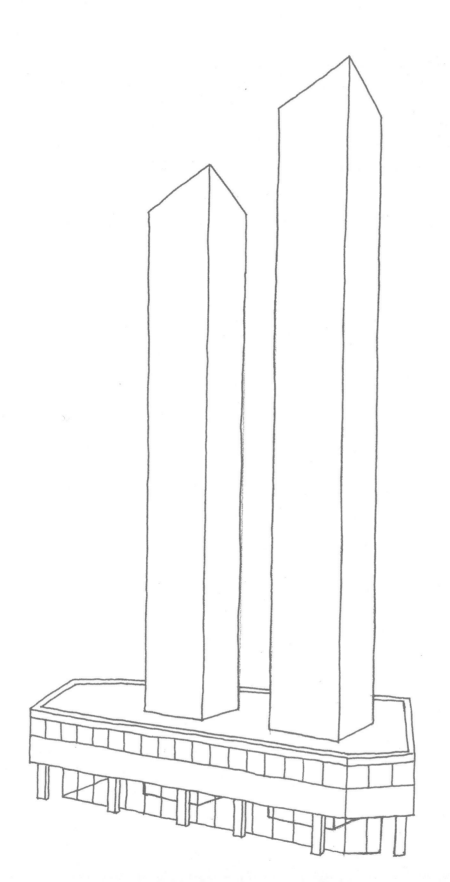

Somerset House,
built in 1776 by Sir William Chambers, is a large neoclassical courtyard building overlooking the River Thames in central London. The buildings have housed a variety of bodies, including government departments and educational societies. More recently, arts and cultural events have taken centre stage. In the last ten years the courtyard has been the venue for a regular outdoor cinema, iceskating, fashion shows, and rock and dance events.

萨默赛特宫是伦敦中心区一个可以远眺泰晤士河的大型新古典主义庭院式建筑。该建筑由威廉·钱伯斯设计，于1776年建成。该建筑容纳各种各样的机构，包括政府部门和教育团体。最近以来，艺术与文学活动已经占据了舞台的中心位置。在过去的十年间，这个庭院一直被作为正规的露天电影、滑冰、时装秀和摇滚活动的举办场所。

你打算在这个庭院中举办什么活动,并且需要什么样的设施来支撑这些活动?

Public toilets
serve a very important function for the citizens of any city. Generally they are pretty ugly, but they don't have to be. The two examples shown here are sculptural and more like art installations set in the environment of a public park.

公共卫生间对于任何城市的市民来说都是一个非常重要的服务设施。通常来说，它们很令人厌恶，但它们并非一定要如此。这里展示的两个例子看上去像是雕塑，同时也更像是设置在公园绿地环境中的艺术品。

金华建筑公园，中国。
DnA建筑与设计事务所，2004年。

步道卫生间，奥斯汀，得克萨斯，美国。
米罗·里维拉建筑事务所，2007年。

设计你自己的公共卫生间，你需要考虑光照、通风和建筑外观。

The **threshold** of a building – its entrance – can be envisaged as the portal between the public realm of the street and the private world of the interior. Most entrances are fairly anonymous and don't really say anything about the people, events and functions inside. The two entrances shown here suggest a narrative beyond the threshold.

建筑的大门——入口——可以被视作街道公共领域和内部私人世界之间的门户。大多数的入口都是相当孤立的，不传达内部人员、活动和功能的任何信息。这里展示的两个入口案例都具有大门功能以外的叙事性特征。

双筒望远镜大厦（前身为奇特/戴大厦），洛杉矶，加利福尼亚，美国。克莱斯·奥登伯格，库斯杰·范·布鲁根和弗兰克·盖里，1991年。

地狱歌舞表演剧场，巴黎，法国，1898年。

为下列某一种功能性建筑设计一个入口：
宠物店、回收中心和巧克力工厂。

When walking through a city we often experience our journey through a **series of framed views** containing places, objects, events, spaces and people. The drawn sequence can be made into a storyboard and is often how film-makers envisage the backdrop to a scene within a movie. My drawings take a walk around the area of St Paul's Cathedral, London. Note that each frame has at least some part of the previous image within it to connect the story together.

我们在城市中穿行时，常常利用一系列包含场所、物体、活动、空间和人员等信息的视框来体验旅程。这种视图序列可以被制作成为一个故事板，也常常被电影制作者用来表达电影中的某个场景。我的这些绘图来源于伦敦圣保罗大教堂附近的城市漫步。注意每一幅画面内都至少含有前一张画面的部分信息，把它连接成一个完整的故事。

将你的城市之旅绘制在系列图板上,注意连接每幅画面。

Man's desire to develop **underwater cities** has in recent years resulted in a number of designers exploring the concept of the 'seascraper'. In this version below, the Malaysian architect Sarly Adre Sarkum envisages a self-sufficient structure that generates its own electricity using solar, wind and wave power. It utilizes its own food-farming system and sustains a small forest on the surface of the vessel. The 'Water-Scraper' is kept upright using a system of ballasts aided by squid-like tentacles that generate kinetic energy.

近年来，人类建设水下城市的愿望引起了众多设计师对"海上大厦"概念的探索。在下图展示的版本中，马来西亚建筑师萨里·阿德尔·萨库姆构想了一个可以利用阳光、风和波动力发电的自给自足建筑。它使用自身的食品-农业系统，并且可以在平台表面维系一片小森林。这个"海上大厦"利用一种乌贼状触须辅助压载系统产生动能保持直立。

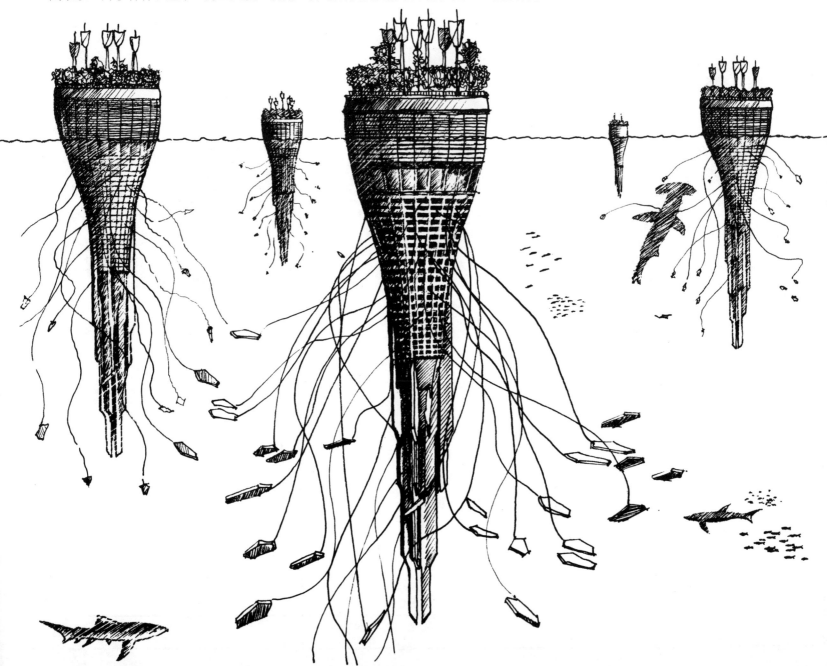

设计你自己的自给自足海上大厦。想想如何产生并获取能量，如何提供食物和生存空间，如何工作和娱乐。

The **Gavina Showroom** in Bologna, Italy, by Carlo Scarpa (1961–1963) is an example of a storefront that defines and frames the display of merchandise using strong geometry, a palette of raw materials such as board-marked concrete, and a Japanese-inspired gate.

位于意大利博洛尼亚的加瓦利亚陈列室，由卡洛·斯卡帕（1961—1963年）设计完成。这是一个使用几何体、横板纹混凝土等粗材料面板，以及日式风格的门突出商品展示区的案例。

设计你自己的展室，同时使用抽象的几何学展示现代艺术作品。

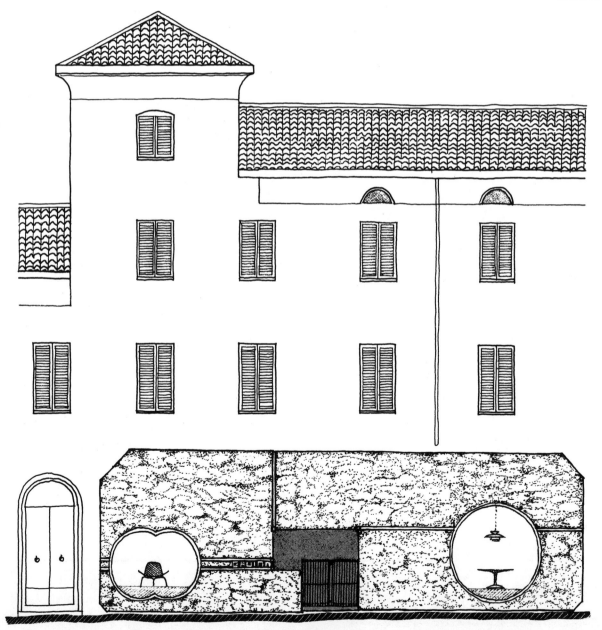

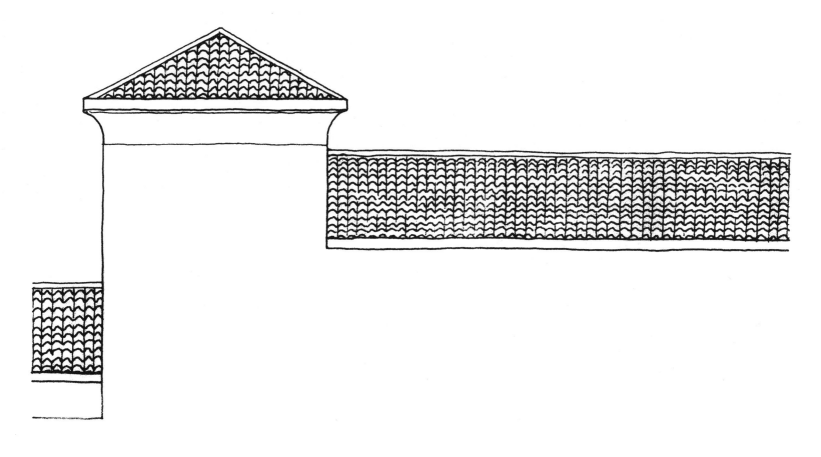

Below is an example of one of the very earliest forms of **natural ventilation** in the city of Hyderabad, founded in 1768 by Mian Ghulam Shah Kalhoro in Pakistan's Sind province. The triangular towers are almost like chimneys on top of the dwellings and are used to funnel cool breezes into the interior. You can also see BedZED, a modern example designed by architect Bill Dunster for environmentally friendly housing on the edge of London in 2002 (below right).

下面的案例是一种最早期的自然通风形式，它位于海德拉巴市，由巴基斯坦信德省的古拉姆于1768年建成。这些住宅顶部的三角塔就像烟囱一样将微风引入室内。你也可以看到贝丁顿生态村，它是建筑师比尔·邓斯特为伦敦市郊的一个环境友好型住房设计的现代案例，该设计完成于2002年（右下图）。

 在这些传统的屋顶上勾勒出一系列集风口的草图想法，同时考虑它们是否能够旋转，以及它们是怎样构成的。

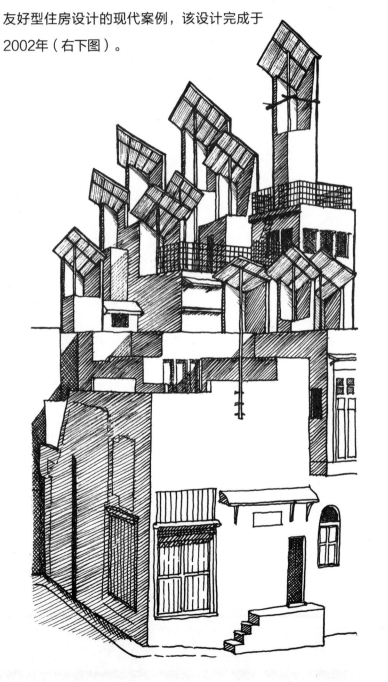

Many cities have a **monument or statue** that symbolizes the freedom or the protection of its people. Rio de Janeiro has a statue of Christ the Redeemer, New York has the Statue of Liberty, and Giza, Egypt, has the Great Sphinx. What does your nearest city have?

许多城市都有一个象征自由或人民保护神的纪念碑或雕塑。里约热内卢有一座救世主基督的雕塑，纽约有自由女神像，埃及吉萨有狮身人面像。离你最近的城市有什么纪念碑或者雕塑？

设计一个你认为能够象征市民保护神或自由精神的雕塑。

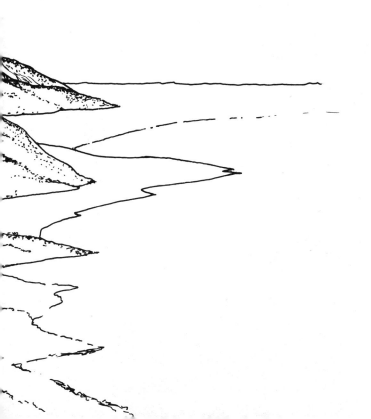

While researching different **bus shelters,** I came across some that exemplified a cultural character and thought it would be interesting to design a bus shelter dedicated to a particular place or city.

在研究不同的公共汽车候车亭时，我遇到过一些能够反映文化特征的案例，并且我认为设计一个属于特定地区或特定城市的公交车候车亭是一件很有趣的事情。

左图：拉瓦卡公交车候车亭，圣安东尼奥，得克萨斯州，美国。卡洛斯·科尔特斯，2005年。

左下图：塔拉兹公交车候车亭，哈萨克斯坦。

下图："瀑布"公交车候车亭，奥兰多，佛罗里达州，美国（John Marhoefer/Walt Geiger），2012年。

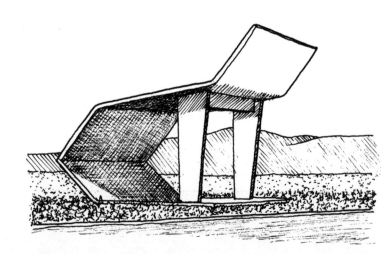

在这个街景中描绘出你的公交车候车亭版本。

Street lighting

Street lighting is a very important part of any city, helping to make spaces safe at night. In recent years street lighting has been developed that mimics how plants open to collect the sun's energy. This example, designed by Philips, opens up and collects solar energy during the day and then closes and emits light at night, but only when people are in close proximity.

街道照明是任何城市中一个非常重要的部分，它有助于在夜间保障空间的安全。近年来，人们已经开发出通过模拟植物开花来收集太阳能的街道照明设施。下面这个案例由飞利浦设计，它可以在白天打开并收集太阳能，同时在夜晚关闭并提供照明，而且它只有在有人靠近时才会提供照明。

勾勒出你自己的可持续街道照明草图，并在你的设计中思考创造可再生能源的方式。

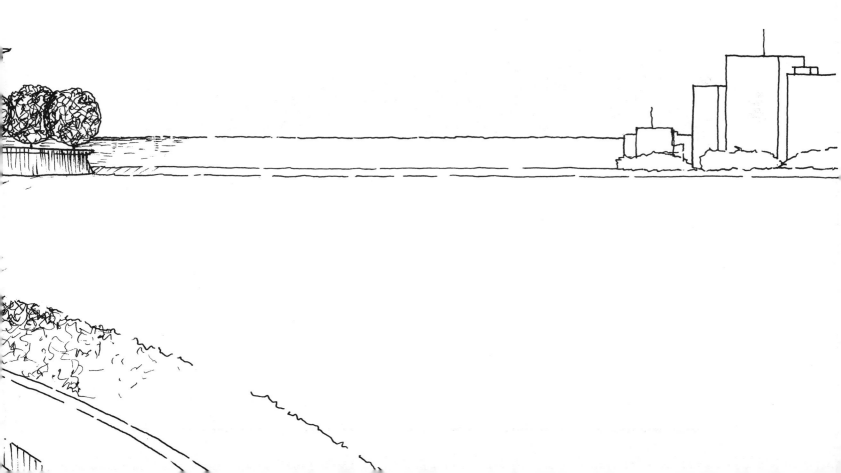

56

Here are some more **city silhouettes.** Try to identify them and then make a new hybrid silhouette using features from each city.

下面是一些城市的剪影。尝试识别出这些剪影所代表的城市，然后利用每个城市的特征来创造一种新的混合剪影。

综合这些城市特征，或者尝试加入你最喜欢的城市地标。

These are copies of the visionary drawings by **Iakov Chernikhov,** the Russian Constructivist architect, completed between 1927 and 1933. He used line, block and plane to express the machine-like quality of his compositions.

这里有俄国建筑师拉科夫·切尔尼科夫于1927~1933年间绘制完成的一些蓝图副本。切尔尼科夫使用线、块体和平面来表达他作品的机械特性。

用透视法延续这个作品，并且在这个城市中加入你自己的延伸内容。

These two examples of Chernikhov's work only use the medium of **line and grid.**
切尔尼科夫的这两个案例只使用了线和网格作为媒介。

在只使用几何线框的情况下画出另外两幅图。

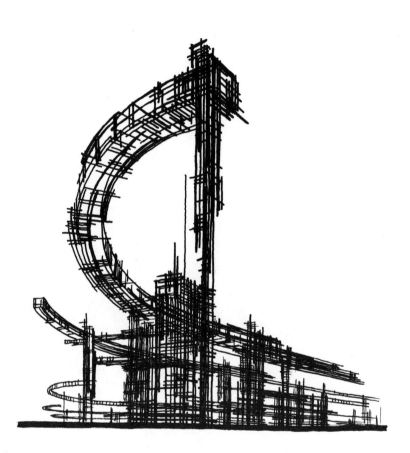

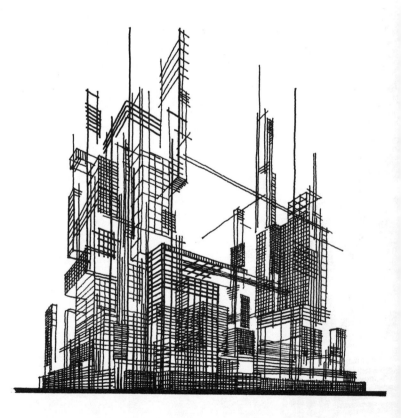

61

These **tram stations** in Hannover, Germany, were designed by Despang Architekten for Expo 2000. While the framework and geometry remain the same throughout the series, the materials used for each individual stop vary, chosen to reflect their local surroundings and local history.

德国汉诺威的这些有轨电车站由德斯庞建筑事务所为2000年世博会设计。虽然整个车站系列框架和几何体保持一致，但每个站点使用的材料都各不相同，这些选取的材料反映了当地的环境和历史。

使用同一种结构创建一系列雨棚，并采用某种方法使每个雨棚的外观有所变化。

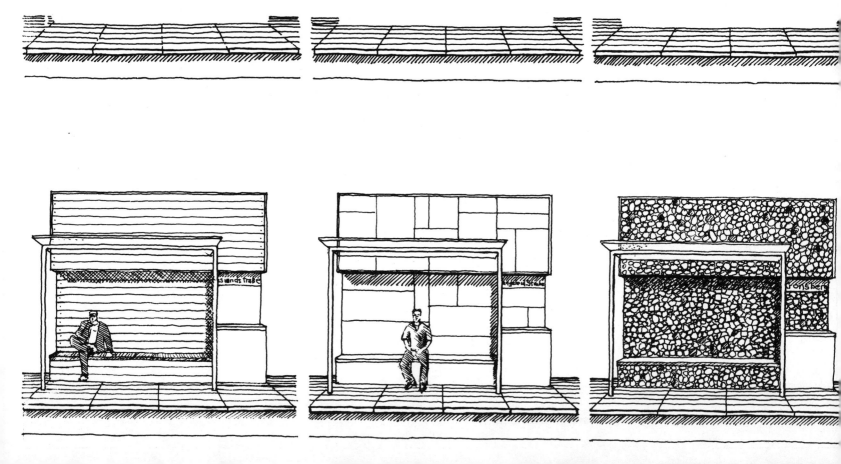

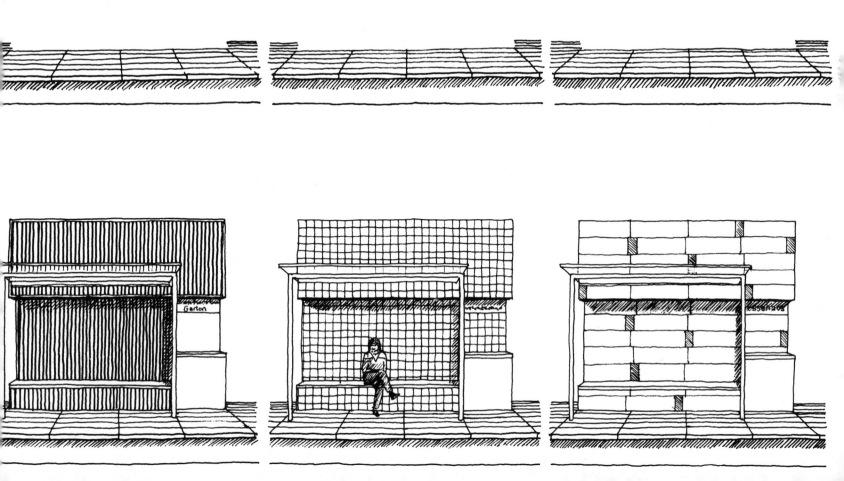

Here is a drawing of the space occupied by **Central Park** in New York City. Design a new park for the twenty-first century, considering the kinds of activities, events and facilities that would engage and enhance the lives of New Yorkers. Would they want sports or art activities? Or maybe a contrast to the city itself, such as nature and wildlife?

这张草图将纽约市中央公园进行了留白设计。请你在中央公园原址上重新设计一个21世纪的新公园,考虑各种有助于丰富并改进纽约人生活品质的活动、事件和设施。市民希望公园中有一些运动或艺术活动吗?或许这个公园将形成与城市的对比,例如更加偏向自然或野生的环境?

在所提供的空间中创造你自己的公园。

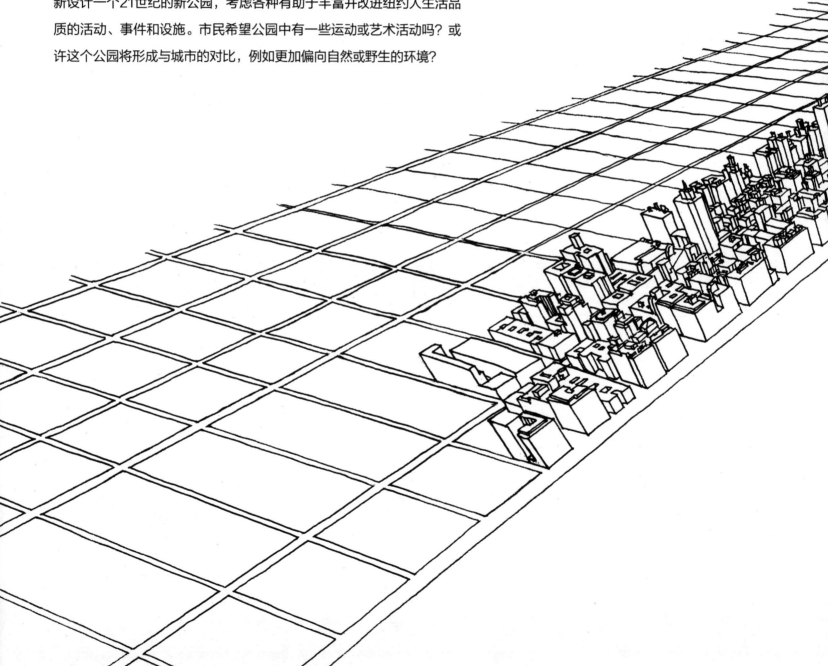

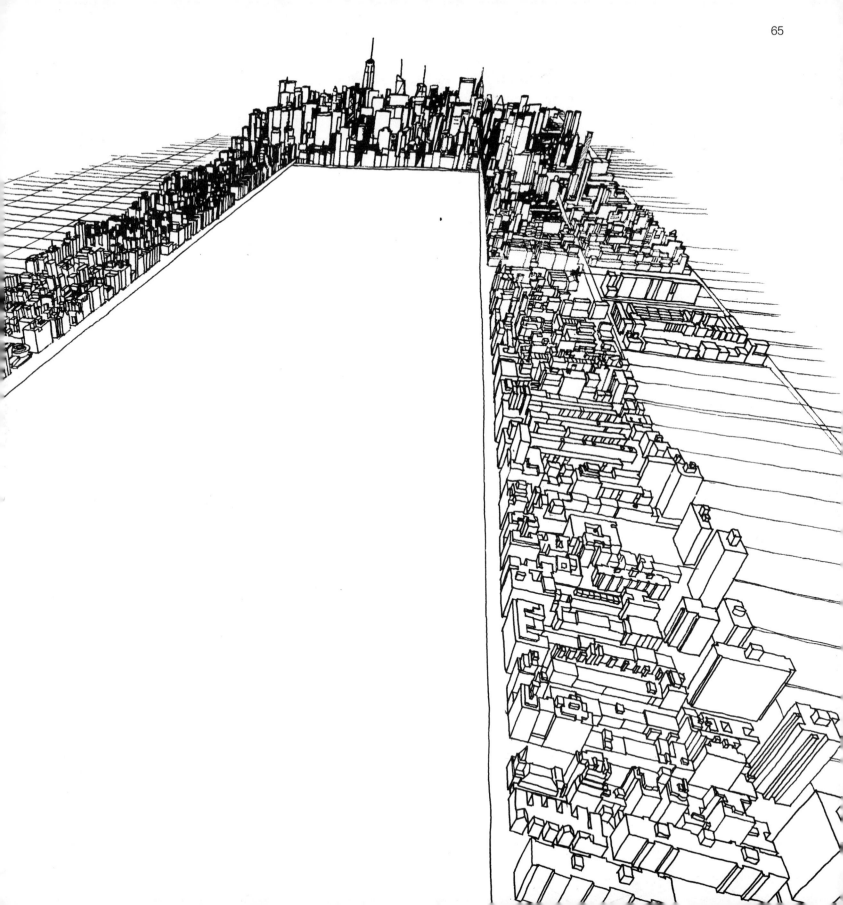

In his book **'Image of the City'** (1960) urban planner Kevin Lynch put forward the idea that cities could be defined using five key elements: paths, edges, districts, nodes and landmarks. To illustrate these concepts I have drawn examples of where these elements can be found in a variety of major cities: London, New York, San Francisco and Barcelona.

城市规划师凯文·林奇在其《城市意象》一书中提出了城市可以由路径、边界、区域、节点和标志五种要素定义的观点。为了说明这些概念，我根据一些主要城市案例绘制了几幅反映这些要素的图像，这些城市包括伦敦、纽约、旧金山、巴塞罗那。

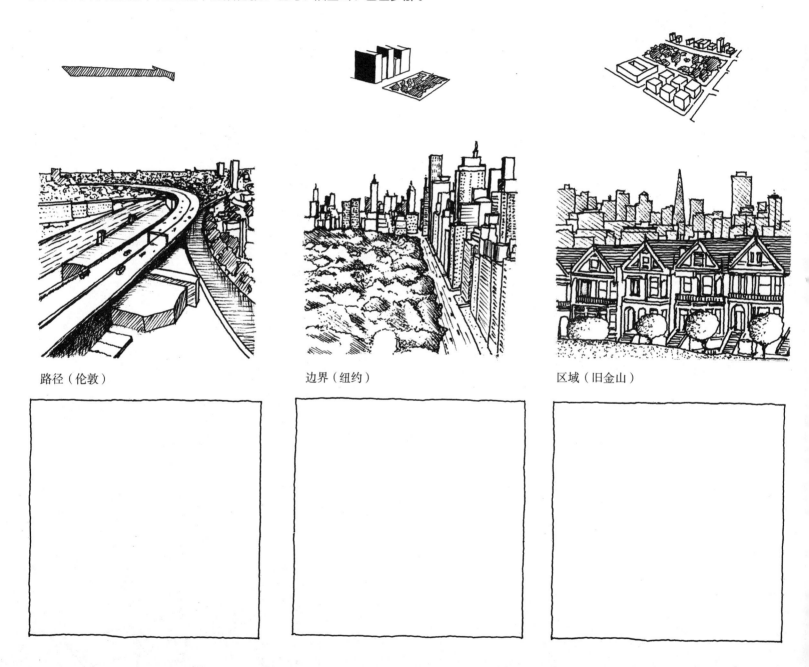

路径（伦敦）　　　　　　　边界（纽约）　　　　　　　区域（旧金山）

在离你最近的城市中走一走,画出可以表示路径、边界、区域、节点和标志的案例草图。

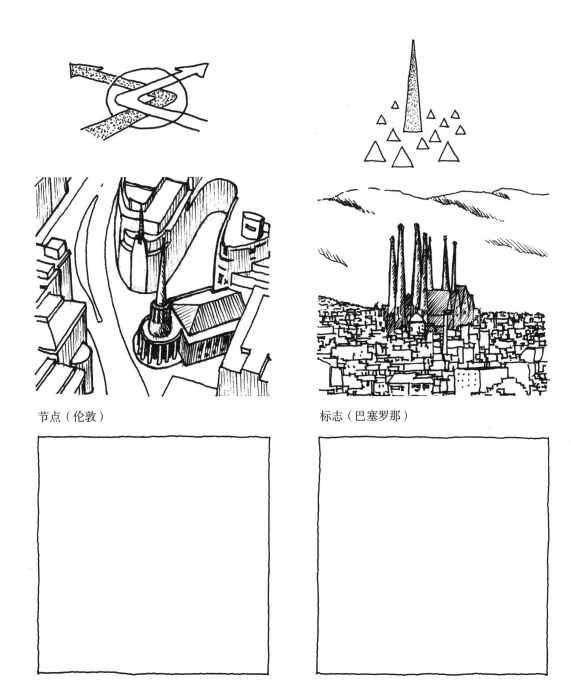

节点(伦敦)　　　　　　　　　标志(巴塞罗那)

Bulgarian artist Christo and Moroccan-born Jeanne-Claude are famous for **wrapping buildings** and landscapes. In 1995 they wrapped the German parliament building (the Reichstag) in Berlin in 100,000 square metres (1.1 million square feet) of fireproof polypropylene and aluminium fabric. The effect was simply stunning.

保加利亚艺术家克里斯多夫和摩洛哥出生的珍妮·克劳德以包装建筑和风景画闻名。1995年，他们用10万平方米的防火聚丙烯和铝制材料将德国柏林的国会大厦（德国联邦议会大厦）包裹起来，效果简直令人震惊。

以这个巴黎的凯旋门草图作为开始，添加你自己的包装效果。

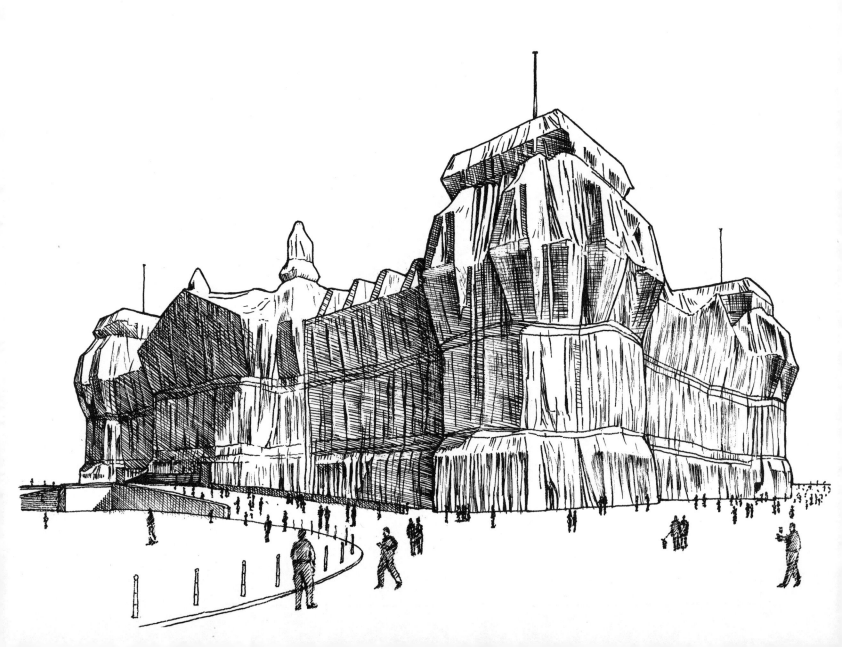

In the 1960s the celebrated avant-garde architectural group **Archigram** challenged conventional ideas about living in cities. The six members of the group designed and contributed design projects in the form of a magazine using pop-art drawings and graphics as a way of exposing their ideas about the future of urban culture. They devised 'Walking Cities', 'Plug-in Cities', and this, my rendering of the nomadic 'Living Pod' (1965–66) by David Greene. Amazingly, these ideas about the relationship between cities and future technology are now 50 years old.

1960年代，著名的先锋建筑小组——建筑电讯学派，挑战了居住在城市中的传统观念。这个六人小组以波普艺术绘画和图像的形式提出自己的设计，并将设计项目投稿在杂志上，借此表达他们对未来城市文化的看法。他们设计了"步行城市"、"插入式城市"以及游牧生活舱，令人惊讶的是，这些有关城市与未来技术的关系的理念已经有50年了。

你将设计什么样的居住舱？
当代技术将如何影响你的设计？ ➡

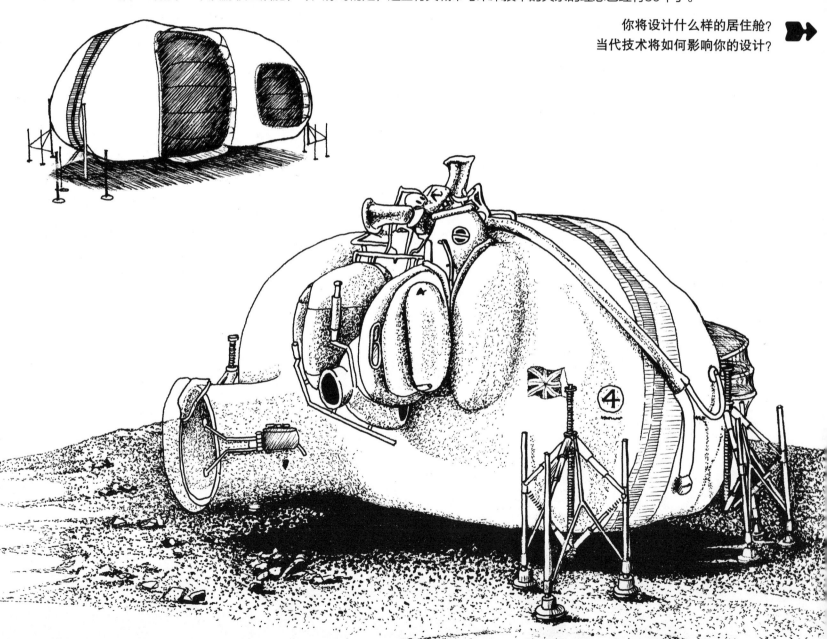

The **shopping arcade** became popular in European cities in the eighteenth and nineteenth centuries and was the predecessor of the North American shopping mall. The arcade provided a safe, pedestrian-only, covered internal street that would provide comfort regardless of climatic and social conditions.

购物拱廊在18~19世纪的欧洲城市十分流行，它是北美购物中心的早期形式。这种拱廊为人们提供了一个安全、纯步行且内部覆盖的街道。且这个街道提供的舒适环境不受气候和社会条件变化的影响。

伯灵顿拱廊，伦敦，萨缪尔·维尔，1819年。

设计你自己的拱廊，你需要考虑拱廊的外形、结构以及通道的照明。你会在其中添加什么样的商店和咖啡厅？

Most **superstore design** is boring. In 1974, the Best Products company challenged this notion by commissioning the Sculpture in the Environment (SITE) architects to design a series of nine buildings across the USA that would re-brand their company. SITE's idea was to take the building envelope and break it down via a series of distortions, fragmentations and displacements.

大多数超市的设计都令人乏味。1974年，百佳产品公司挑战了这一观念，它们委托环境雕塑组织的建筑师设计了横跨全美的一系列9栋建筑，这次行动也重塑了他们的公司形象。环境雕塑组织的想法是将建筑包裹起来，并且通过一系列扭曲、碎片化和重置的手法将其分解。

Notch陈列室，迈阿密，佛罗里达。
SITE，1979年。

Peeling工程陈列室，里士满，弗吉尼亚。
SITE，1972年。

倾斜陈列室，陶森市，马里兰州。
SITE，1976年。

以这个给定的简单盒子作为起点，
你将如何让你的超市与众不同？

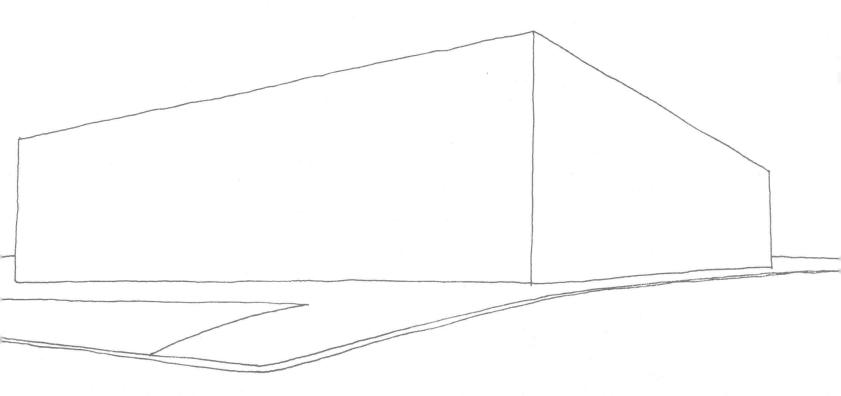

Images of the city have played an important role in many **film, theatre and music** performances. The example below, influenced by the film Metropolis by Fritz Lang (1927) was the design of the stage set for David Bowie's 'Diamond Dogs' tour in 1974.

城市意向在很多电影、喜剧和音乐表演中扮演着重要的角色。下面这个案例，受到了弗里茨·郎的电影《大都会》（1927）的影响，最终成为大卫·鲍伊1974年巡演"钻石狗"的设计舞台场景。

根据某个城市的意向为一个戏剧、音乐或摇滚表演设计你自己的舞台场景。

"钻石狗"舞台布景，朱尔斯·菲塞尔和马克拉维兹设计，1974年。

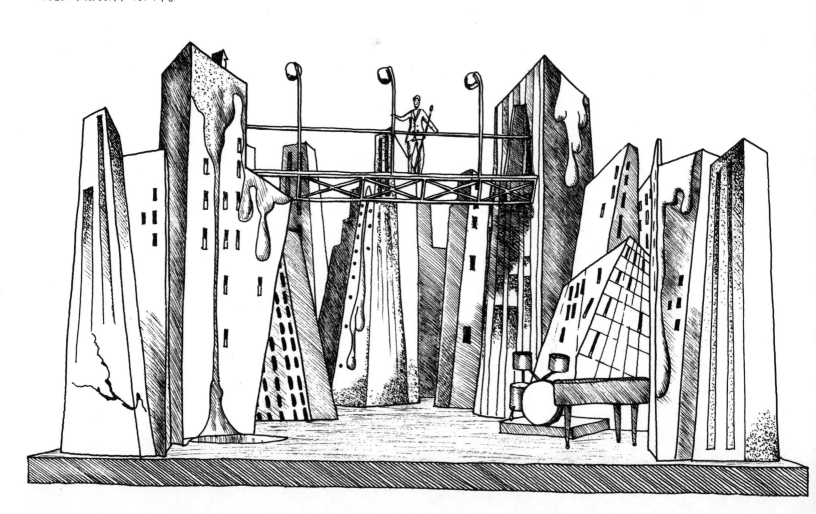

77

The district of Akihabara, Tokyo, is known as **'Electric Town'** because of its shops dedicated to electronic goods and its array of neon signage. Below is a typical street showing signage, advertising and lighting, and on the opposite page is the same street without any signs.

东京的秋叶原地区因密集的电子产品商店和林立的霓虹灯招牌而被称为"电子城"。下图是一个展示招牌、广告和灯光的典型街道,而另一页则是没有任何标志的相同街道。

给下边的图上色并且在右边创造你自己的电子城。

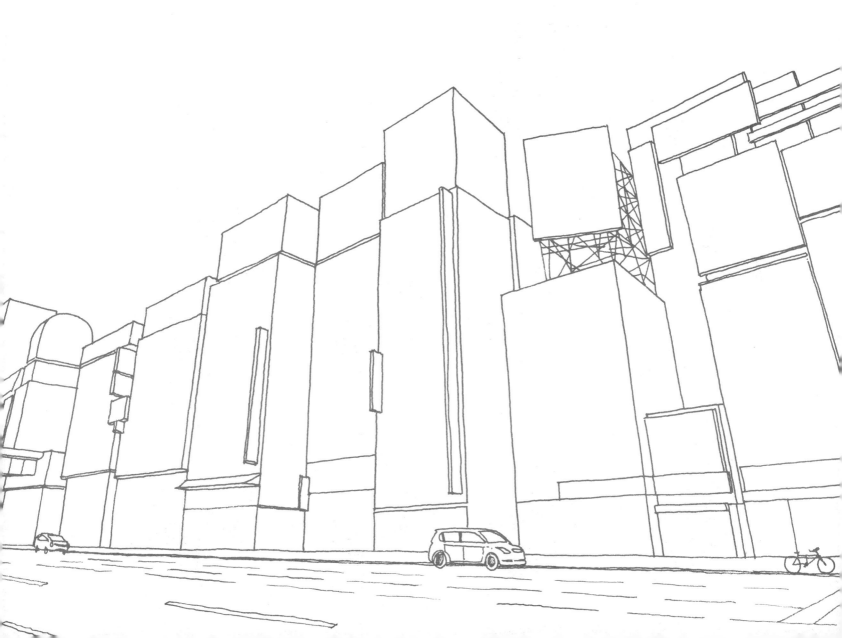

Roosevelt Island

is a narrow island in New York's East River, which has been the site of numerous architectural competitions over the years. One particular competition entry, from the German architect O.M. Ungers in 1975, imagined the city development as a series of blocks formed from different typologies, such as towers, courtyards and terraces.

罗斯福岛是纽约东河的一个狭窄岛屿，多年来这里一直是众多建筑竞赛的所在地。其中一个项目是1975年德国建筑师O·M·昂格尔的作品，该项目将城市想象成由塔、庭院和平台等不同原型构成的一系列街区。

你还可以在这些昂格尔街区中设计出其他什么变体么？

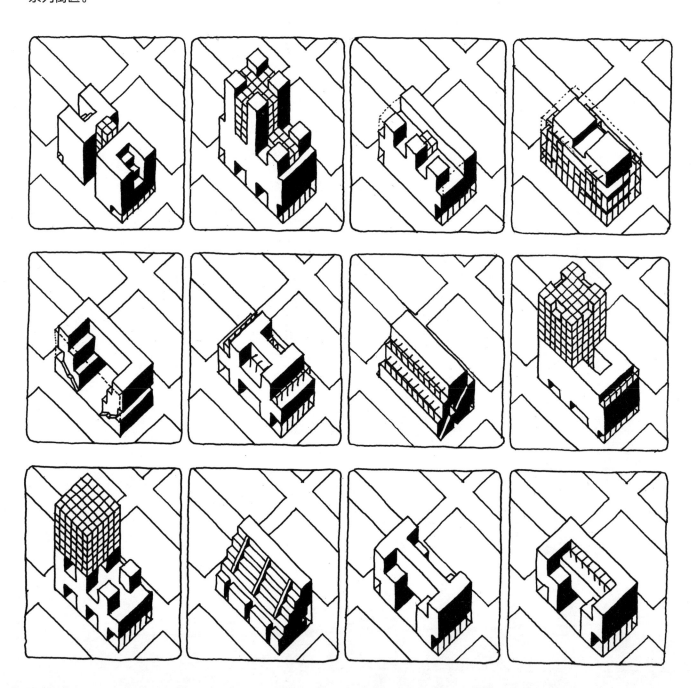

The corners and junctions of streets are often celebrated by important **landmark buildings.**
The Lois & Richard Rosenthal Center for Contemporary Art, Cincinnati, USA, by Zaha Hadid Architects (2003) deals with the variety of conflicting scales of the surrounding buildings by using a variety of horizontal forms.

街道的拐角或者交叉口常常因重要地标性建筑而增光添彩。位于美国辛辛那提的洛伊斯和理查德·罗森塔尔当代艺术中心，由扎哈·哈迪德建筑事务所于2003年设计完成，它通过使用各种水平形式来处理周围建筑物的各种冲突的尺度。

考虑到同样的情况，你会使用什么样的3D形式、结构或者尺度来创造一个成功的街角建筑？

Famed for inventing the geodesic dome, US architect, engineer and inventor **Buckminster Fuller** proposed putting a vast dome over the whole of Manhattan Island, New York. He also calculated that given the right atmospheric conditions, such as in the heat of the desert, his geodesic sphere would float because the heated air within the structural tubes would be lighter than the surrounding ambient temperature.

因发明"网格球顶"而闻名的美国著名建筑师、工程师和发明家巴克敏斯特·富勒建议在整个曼哈顿岛地区放置一个大的圆顶。

想象一个漂浮的城市位于富勒网格球顶内，并且在提供的空间内将其描绘出来。

This is a typical **public space** in a city shopping area. What facilities would you provide to make this an interesting and usable space?

这是一个典型的城市购物区内的公共空间。你会提供什么样的设施将它变成一个有趣且有用的空间?

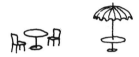

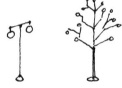

试着添加树、咖啡桌、亭子、照明设备、喷泉、雕塑、长椅或者滑板坡道。

In 2011 a competition to design a new **metro entrance** in the centre of the city was launched in San Sebastián, northern Spain. These examples from architects Snøhetta (first prize) and BABELstudio show different approaches to shelter, threshold and materials.

2011年，西班牙北部的圣巴斯蒂安举办了一个在城市中心设计一个新地铁入口的竞赛。这些来自建筑师斯诺赫塔事务所（第一名获得者）和巴别塔工作室的案例展示了雨棚、入口和材料的不同设计方法。

针对一个新的地铁站点构思出自己的草图想法，同时考虑周边的街景。

上图：巴别塔工作室
下图：斯诺赫塔

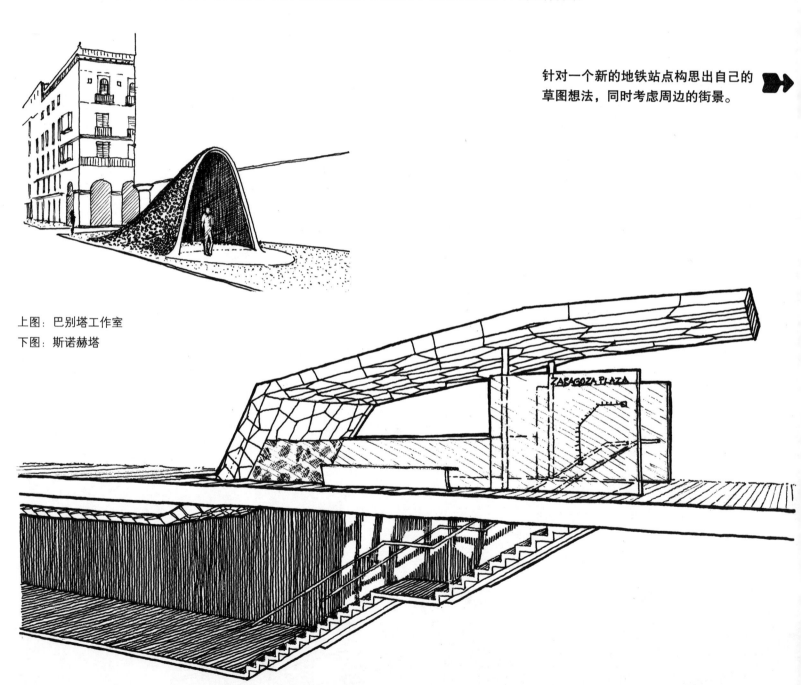

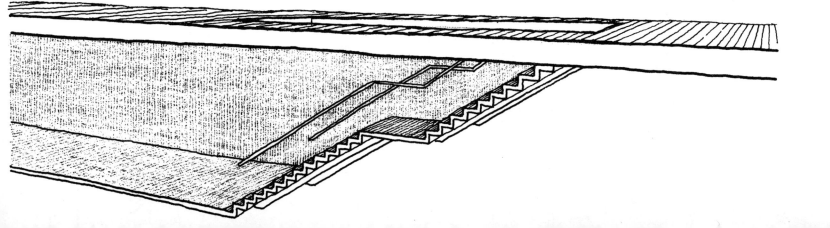

High-density urban areas characterize many
Mediterranean towns and cities, with people living in buildings tightly packed together. Issues such as privacy, noise, light, views and outside space are a major concern for the inhabitants. Preserving the basic form of the buildings, how would you alter their relationship to each other?

许多地中海城镇都呈现出高密度城市地区的特征，在这些城市里人们始终保持高度拥挤状态。隐私、噪声、光线、景观和户外空间等成为居民们的主要关心问题。保留这些建筑的基本形式，你会如何调整他们之间的彼此关系？

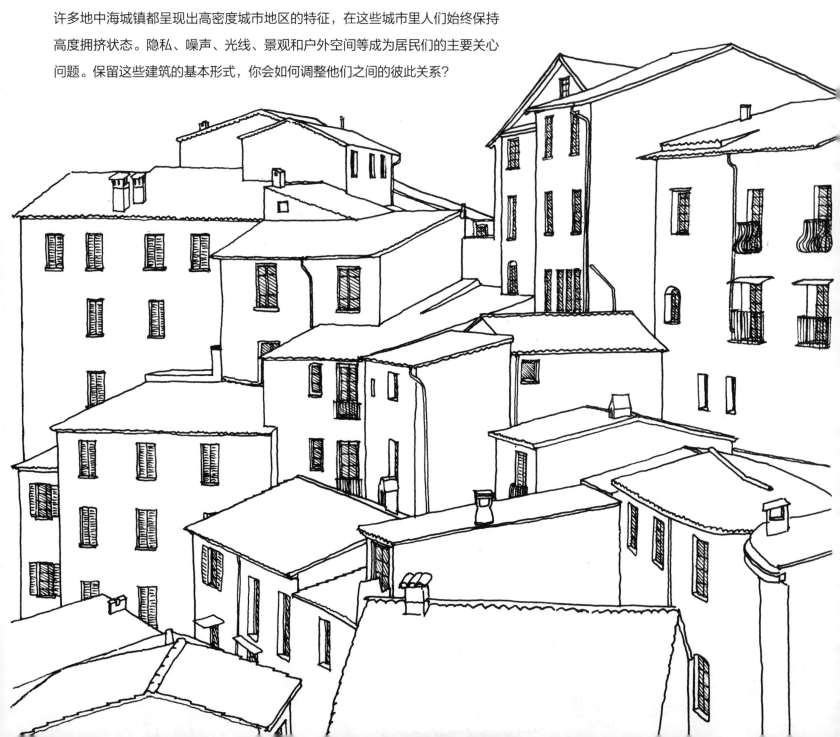

设计新型的窗户和阳台，扩建部分以及不同类型材质——你也可以在房屋间架设廊桥。

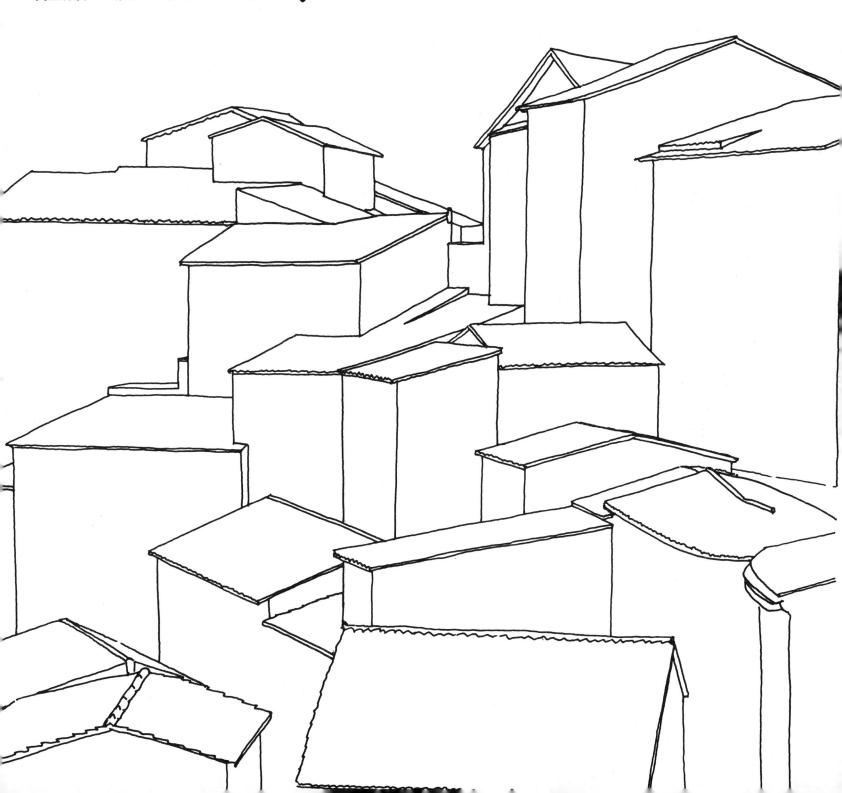

In her designs for a series of buildings for the **Sugamo Shinkin Bank** in Tokyo, the French-born architect Emmanuelle Moureaux challenged the image of the bank as a building that is predominantly formal and staid. Her clever use of geometry, massing, colour and manipulation of window openings offered a new freedom to a building type traditionally known for its sombre appearance.

在为东京的巢鸭信用银行设计的系列建筑作品中，出生于法国的建筑师埃马纽埃尔·穆罗挑战了银行建筑一贯以来的正式和刻板形象。她巧妙地运用几何学、体块、颜色以及窗口控制，解放了原本以阴沉外观而为人所知的建筑类型。

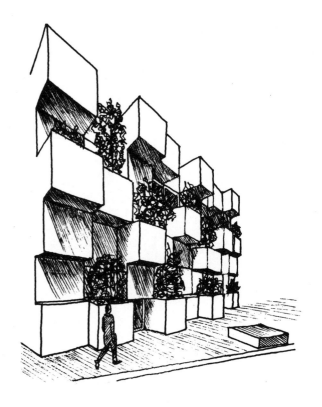

巢鸭信用银行，长崎支行，东京，2014年。

巢鸭信用银行，志村支行，东京，2013年。

巢鸭信用银行，常盘台，东京，2010年。

为其他需要在21世纪进行大改造的机构构思草图,也许从你当地的银行开始。

In 1922 the American architect Hugh Ferriss made a series of drawings detailing **'the cutback principle'.** These demonstrated how skyscrapers in major American cities could be modelled to allow light into the surrounding streets. The cutback principle had formed the basis for zoning laws introduced in 1916, and Ferriss illustrated what high-rise structures might look like based on this idea.

1922年，美国建筑师休·费里斯绘制了一系列详述"切削原则"的图画。这些图画展示出了美国大城市中的摩天大楼为了满足周围街道采光的要求而形成的体块效果。切削原则形成了1916年区划法的基础。费里斯以图示的形式展示了基于这种概念的超高层建筑效果。

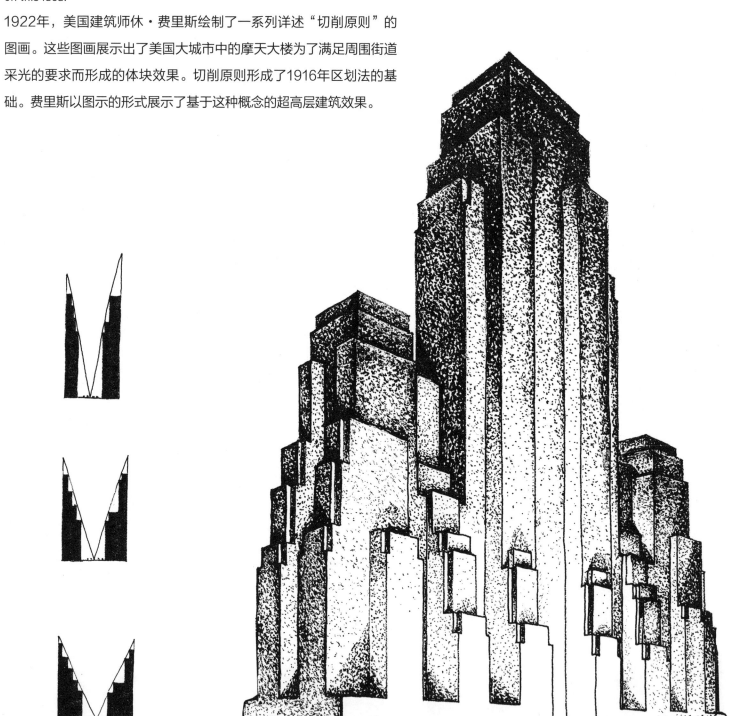

以这个部分成形的建筑体块为起点，使用切削原则设计你自己的摩天大楼。

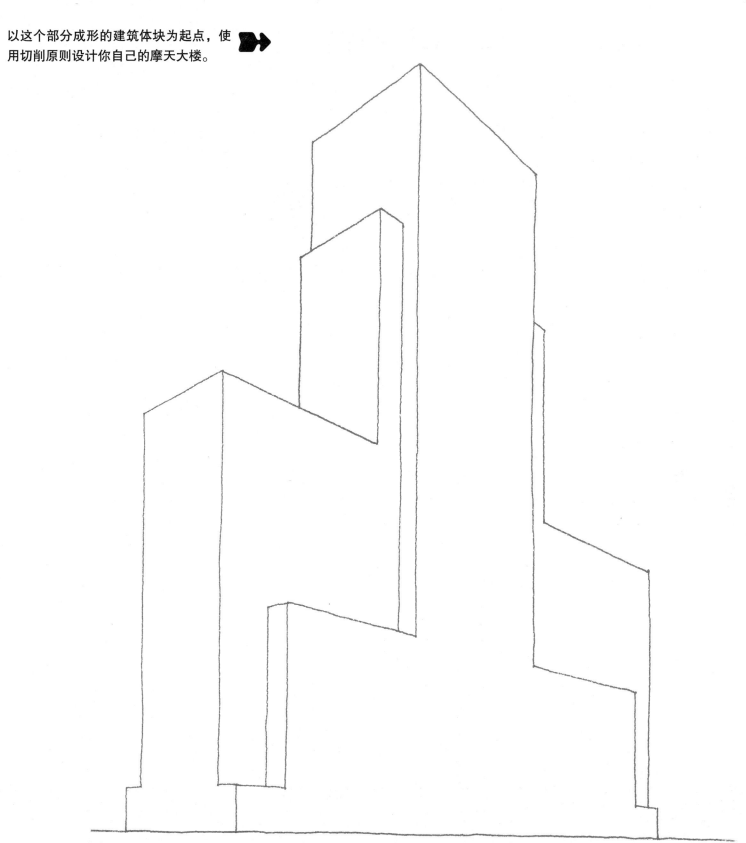

Entitled 'Very Large Structure', this project for a **nomadic city** is the vision of Spanish architect Manuel Domínguez. Similar to Ron Herron's (Archigram) 'Walking City', VLS moves on massive caterpillar tracks across desert terrains. Add your own city to this partially completed drawing.

西班牙建筑师曼纽尔−多明戈斯设计的游牧城市项目被称为"超大建筑"。与隆·赫伦（建筑电讯学派）的"步行城市"类似，超大建筑可以依靠大型履带在沙漠地带移动。将你的城市补充到这个尚未完成的画作中。

设想处在一个十分艰苦的环境中，我们需要什么样的建筑来维持生命。

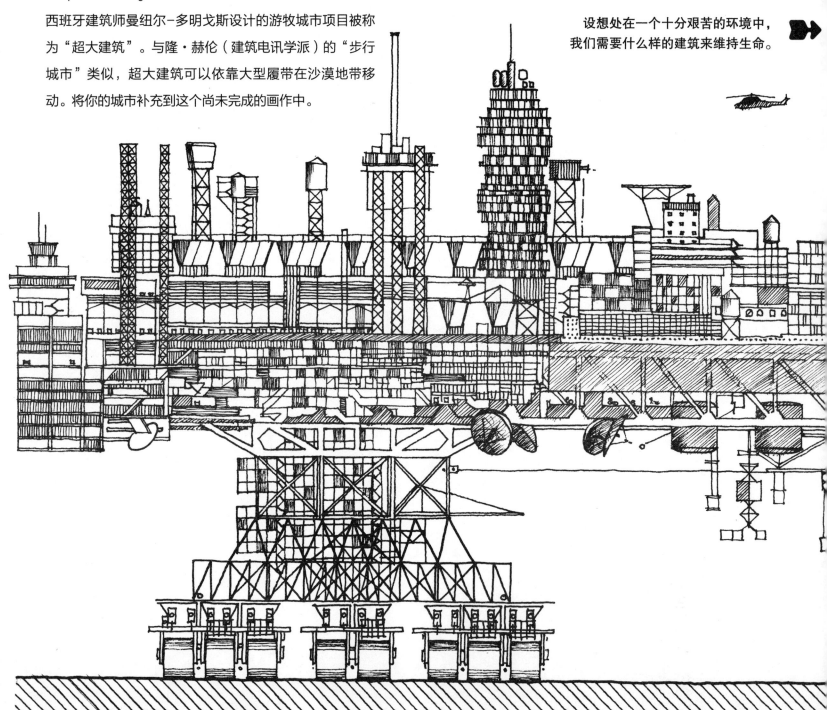

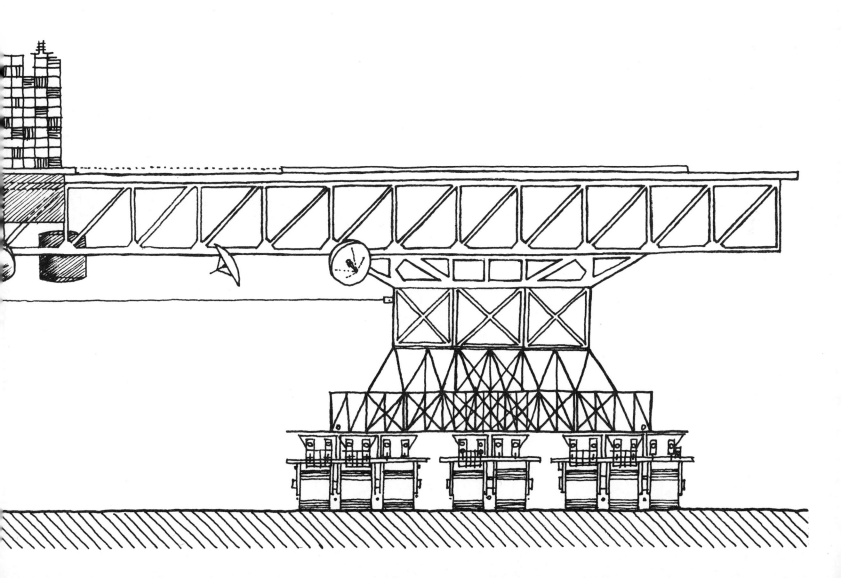

Airport buildings are often the first and last memory that one has of a city when visiting a foreign country. The TWA terminal at JFK International Airport, New York, was designed by Eero Saarinen and Associates in 1962 and was based on the idea of a large bird either landing or about to take off.

机场建筑常常是人们去国外访问时对某个城市的最初记忆，同时也是最终的记忆。纽约肯尼迪国际机场的TWA航站楼由埃罗·沙里宁联合事务所于1962年设计完成，该设计灵感来源于一个着陆中或即将起飞的大鸟。

为离你最近的城市机场构思一个方案草图，它有什么特征，它是否与国家的特色、文化和区位有关？

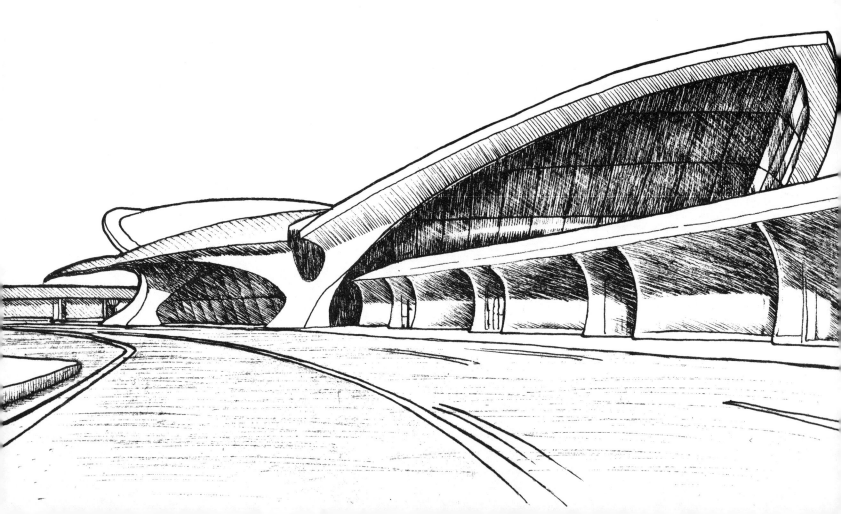

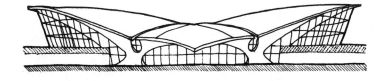

This **maze** is a partially completed design for a labyrinthine city. One side faces the ocean and the other side faces the forest. Imagine that you have arrived by boat and have to negotiate a path through the city to reach your destination in the forest.

该迷宫是为一个迷宫般的城市所做的部分设计。它一面临海，另一面靠近森林。想象一下，你刚刚乘船到达，必须找到一条穿越城市并到达对面森林中目的地的道路。

完成这幅图并给其上色，加入住宅和花园。

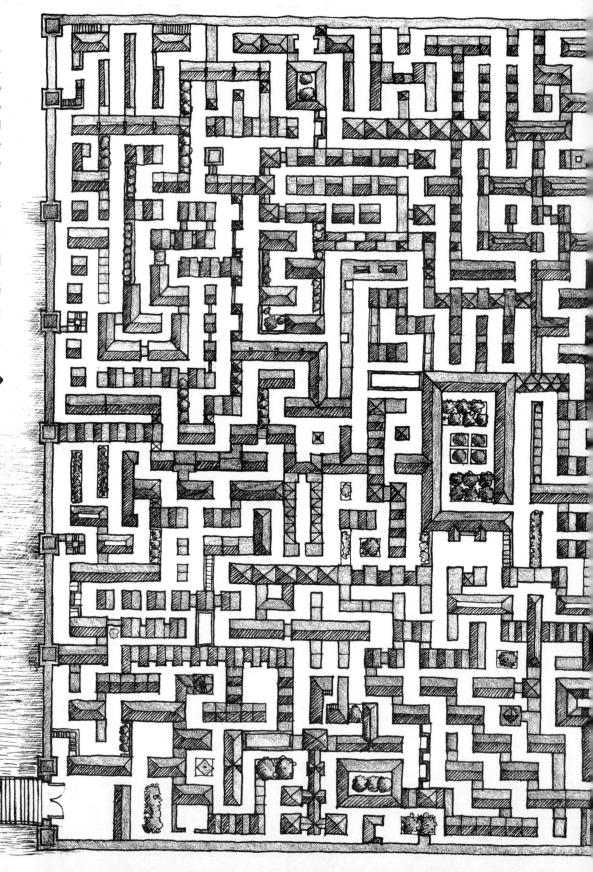

In the 1950s and 1960s at **service stations** on the Italian Autostrade, the Pavesi Autogrill restaurants celebrated the luxury of travelling by car by constructing heroic bridge buildings spanning the roads. Their customers could relax while enjoying the latest automobile fashion parade.

在20世纪50年代和60年代的意大利高速公路服务站上，帕韦西·奥托格里尔餐馆通过建设史诗般的跨路桥建筑庆祝当时属于奢侈行为的汽车旅行。他们的顾客在享受最新的汽车时尚游行的同时，放松身心。

借鉴这个意大利案例的思想精髓设计一个新的桥梁建筑。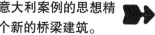

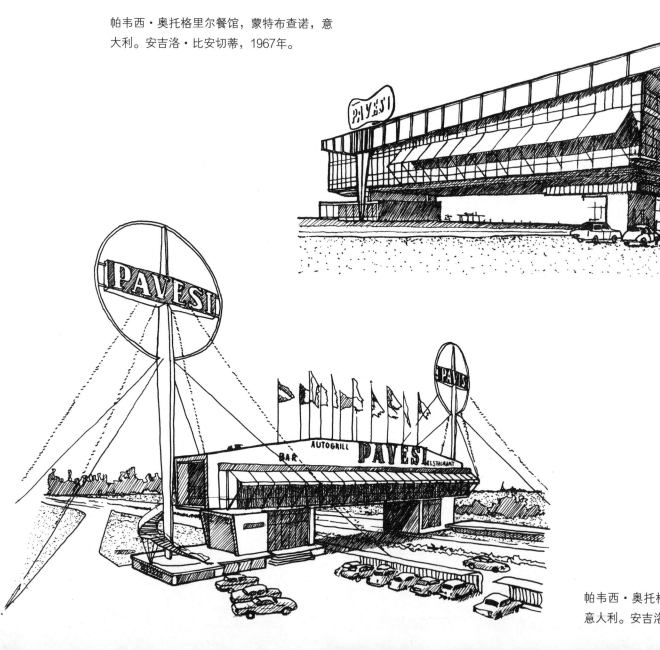

帕韦西·奥托格里尔餐馆，蒙特布查诺，意大利。安吉洛·比安切蒂，1967年。

帕韦西·奥托格里尔餐馆，费伦佐拉亚达，意大利。安吉洛·比安切蒂，1959年。

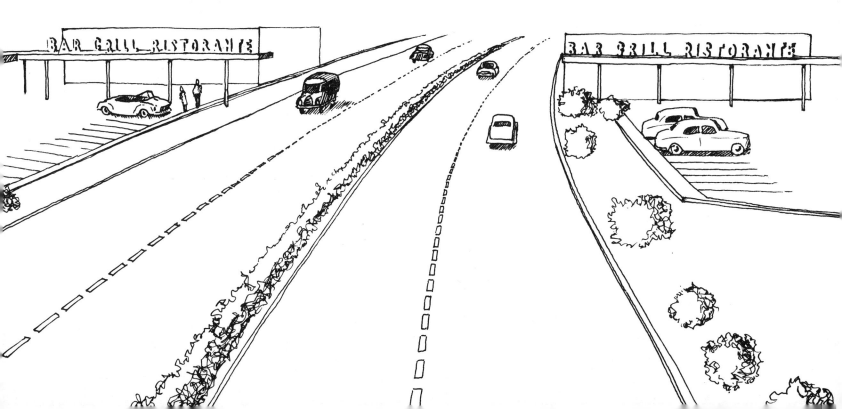

These **housing typology diagrams** were taken from one of my old sketchbooks and compare how one might organize arrangements for types of housing in the city. The generic types are terraces, courtyards, blocks, towers and villas, with each one a variation on the next.

我从一本旧的写生书中找出来一些住宅类型图，并对比了城市中房屋类型的组织安排。通常的住宅类型有阶梯式、院落式、街区式、塔式和别墅，每一种都较其他有所变化。

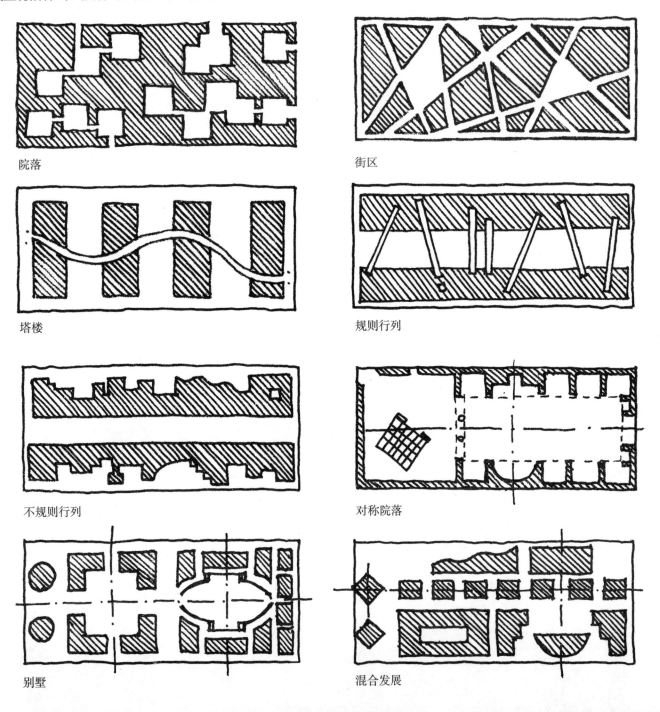

院落

街区

塔楼

规则行列

不规则行列

对称院落

别墅

混合发展

绘制出你自己的版本,考虑尺度、比例,空间分布以及街道和花园的采光。

Gas stations

are one type of building that appears in almost every city and yet does not feature in many architectural journals. Often viewed as a piece of background architecture with functional design ambitions, the gas station has occasionally given rise to some expressive structures. The ones illustrated have become landmarks within the urban landscape.

加油站是一种几乎所有城市都有，却并不常出现在建筑杂志中的建筑类型。加油站通常被认为是一种具有功能性的背景建筑，但是偶尔也会出现一些有表现力的结构。这些插图中出现的案例已经成为这些城市景观中的地标。

"汽车宫殿"加油站，穆尔德斯韦克，尼梅根，荷兰。约翰·米尔曼和约翰·凡·德·皮吉尔，1936年。

恩佐加油站，棕榈泉，加利福尼亚州，美国。艾伯特·弗雷和罗布森·钱伯斯，1965年。

绘制一个服务站的草图方案，使它能够成为你所在城市的一个标志。

NP gas station, Madrid, Spain. Moneo-Brock, 2007.

In his 1910 essay 'Ornament and Crime' the Austrian architect **Adolph Loos** put forward the idea that superfluous decoration on buildings was akin to a criminal act. Loos' design for the Müller House predicated the Modernist, Functionalist and, more recently, Minimalist movements of the 20th century, which had a huge impact on the appearance of our cities. More recently, though, digital manufacturing techniques have resulted in buildings becoming more decorative again. How might this change the appearance of the Müller House?

奥地利建筑师阿道夫·卢斯在他1910年的论文"装饰和罪恶"中提出一个观点——"装饰即罪恶"。卢斯设计的穆勒住宅基于现代主义、功能主义和20世纪兴起的极简主义运动，后者对城市的外观产生了巨大的影响。然而近来数字制造技术再次使建筑变得具有装饰性。这将如何改变穆勒住宅的外观呢？

通过装饰穆勒住宅创造你自己的"罪恶"。

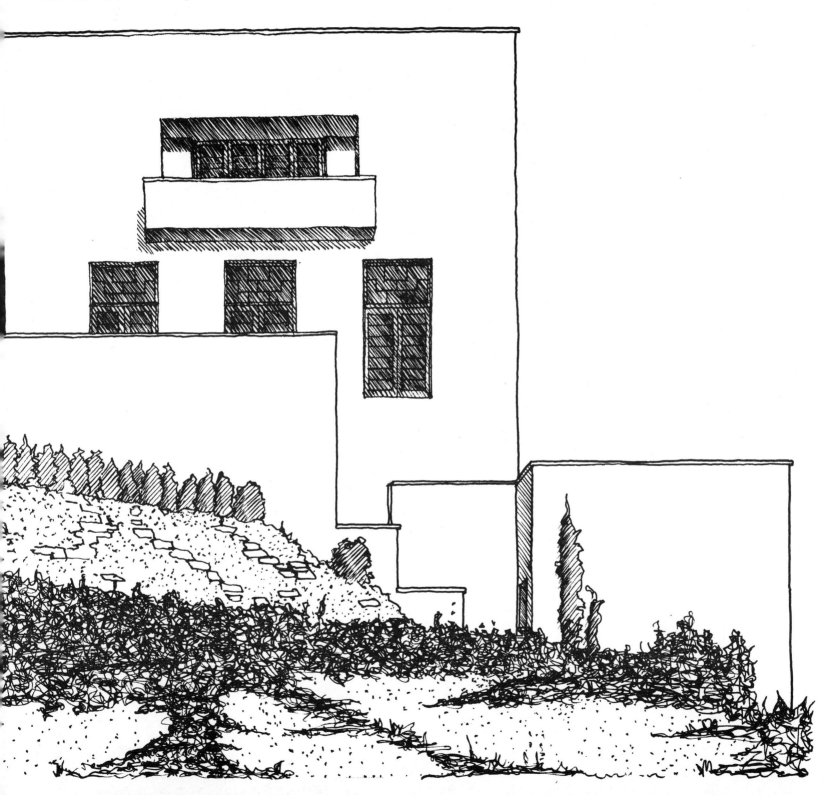

The Dynamic Tower, also known as the Da Vinci Tower, is a proposal for a mixed-use moving skyscraper designed by architect David Fisher for Dubai. Each floor rotates independently around a fixed core and it is proposed that the whole building will be powered by wind turbines and solar panels, with the major components being prefabricated off site.

动力塔，又被称作达·芬奇塔，是建筑大师大卫·费希尔为迪拜混合动力摩天大厦提供的设计方案。这栋大楼的每一层都围绕固定的核心独立旋转，并期望整栋大楼由风力涡轮机和太阳能电池板提供动力、同时设想主要部分均采用预制构件。

如果你要设计一座摩天大楼，他将是什么样子？考虑一下你可能获取并供给动能的方法。

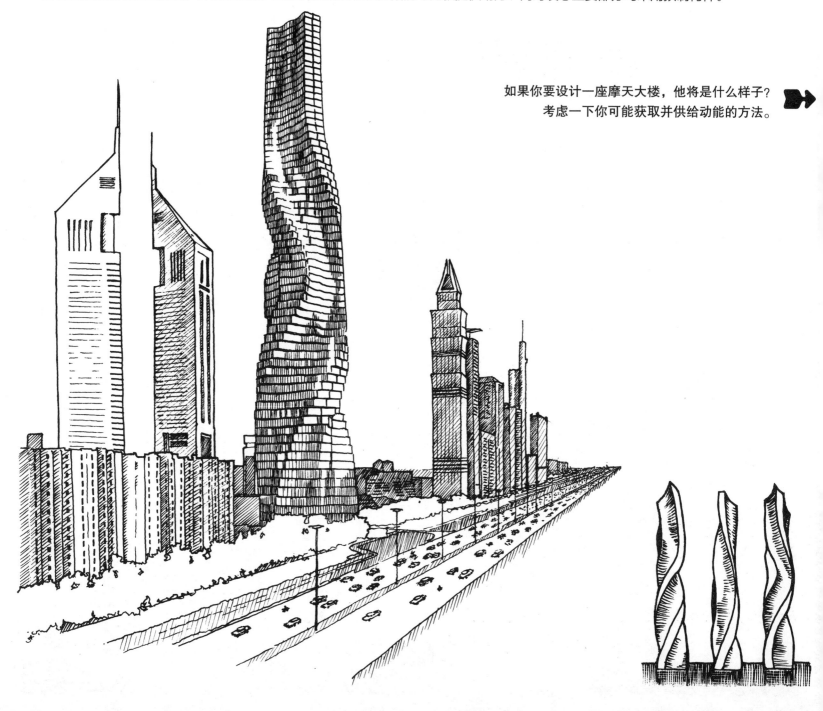

The **Nolli Plan** of Rome was surveyed and drawn between 1736 and 1748 by the Italian architect Giambattista Nolli (sometimes known as Giovanni Battista). His maps are significant because they describe the interiors of public buildings as if they were exterior spaces like streets and squares – in other words, public spaces. On the right-hand page I have erased a section of this map for you to complete.

罗马的诺利平面由意大利建筑师詹巴蒂斯塔·诺利（有时被称为乔万尼·巴蒂斯塔）于1736～1748年间测量并绘制完成。他的这些示意图纸意义重大，因为它们像描绘街道和广场等外部空间（像公共空间）一样描绘公共建筑的内部空间。我抹去了右边页面中的部分图纸，这些图纸将留待你去完成。

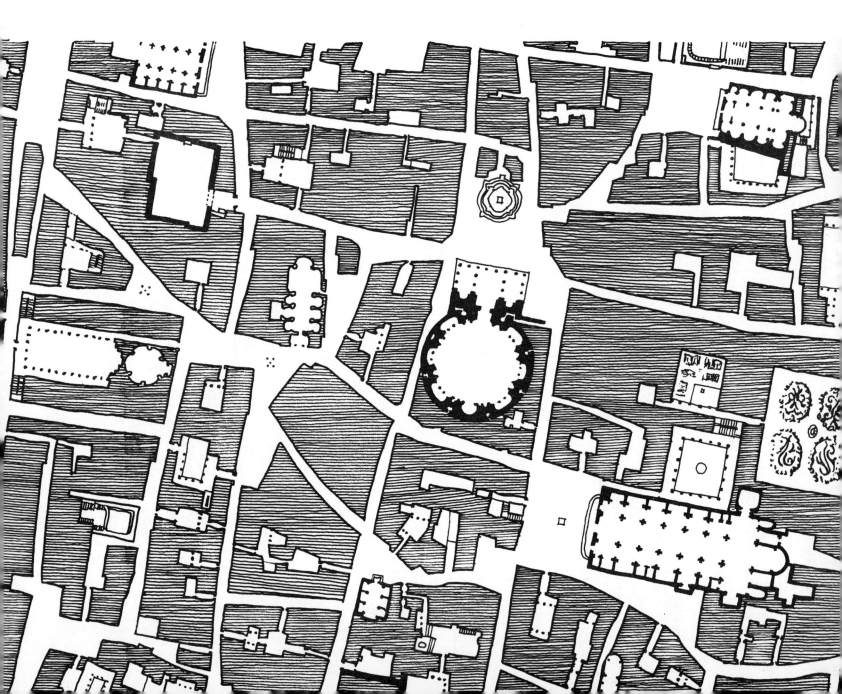

绘制庭院、广场、街道、拱廊、图书馆、教堂、清真寺、公园或者花园。

Here we can see **artists interacting with architecture.**
The first artist attached an inflated bubble containing a palm tree and hammock to a museum; the second cut an eight-metre- (six-foot-) diameter disc into the facade of a local building and attached it to a motor, turning this section of the building inside out in a cycle lasting just over two minutes.

这些图纸反映了艺术家与建筑之间的互动。第一个艺术家在博物馆上附加了一个内含棕榈树和吊床的膨胀气泡。第二个艺术家在一个本地建筑的正立面上切割出一个8米直径的圆盘,并且将其连接到电动机上。这部分建筑可以在2分多钟的时间内进行内外翻转循环。

提出一个能够实现建筑内外互动的装置。它是否涉及切割、掩饰、剥离或叠加?

气候胶囊,绿洲7号,工艺美术馆,德国汉堡。豪斯拉克科,1972年。

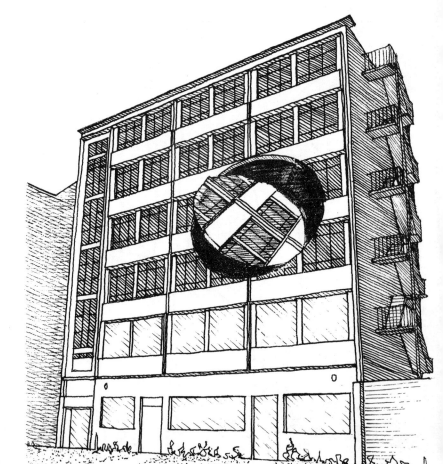

"会转动的大厦",利物浦双年展,理查德·威尔逊,2007年。

The proportions of a street — the general building height compared to the width — will determine how much daylight illuminates the space. This will invariably affect the mood of the place. As you can see from these sectional street diagrams, the proportion is determined by height to width and is measured in imaginary squares (ratios such as 1:1, 1:2, 1:3 or even 1:1.5).

街道高宽比（一般建筑高度与街道宽度的比值）决定了街道内的空间采光亮量，也一定影响着场所的气氛。正如你从这些街道剖面图中所见，这个比例取决于高度和宽度比值，且测量于虚拟的直角空间内。（比例如1：1，1：2，1：3甚至1：1.5）

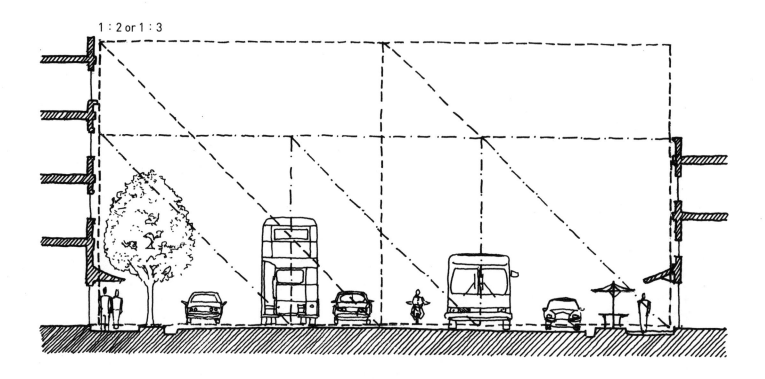

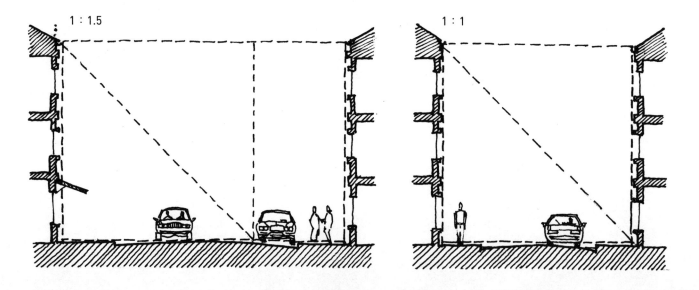

勾勒出你的街道草图并且尝试确定街道高宽比。

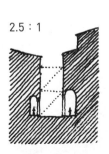

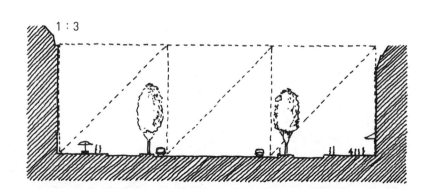

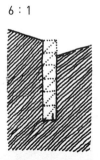

The centre of London has many beautiful **public squares** but the gardens in the centre of them are often private. The example shown is Bedford Square in Bloomsbury, London.

伦敦市中心有很多美丽的公共广场，但是广场中的花园往往是私有的。这里展示的例子是伦敦布卢姆斯伯里的贝德福德广场。

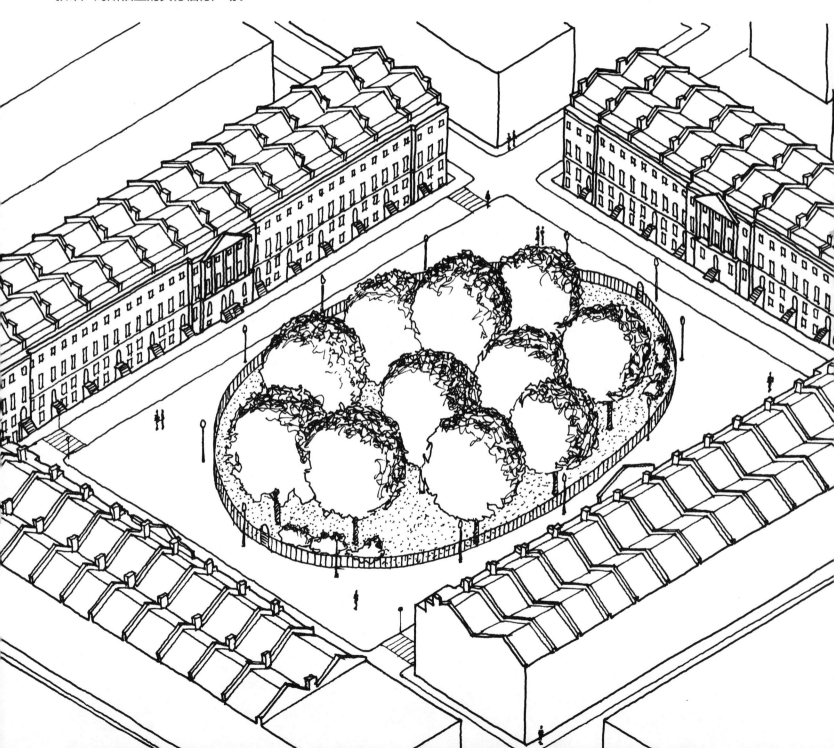

想象一下如果这个中心花园向公众开放。你将会保持公园现状,还是建议一个不同的景观?谁将会使用它呢?是办公室职员,购物者还是孩子们?

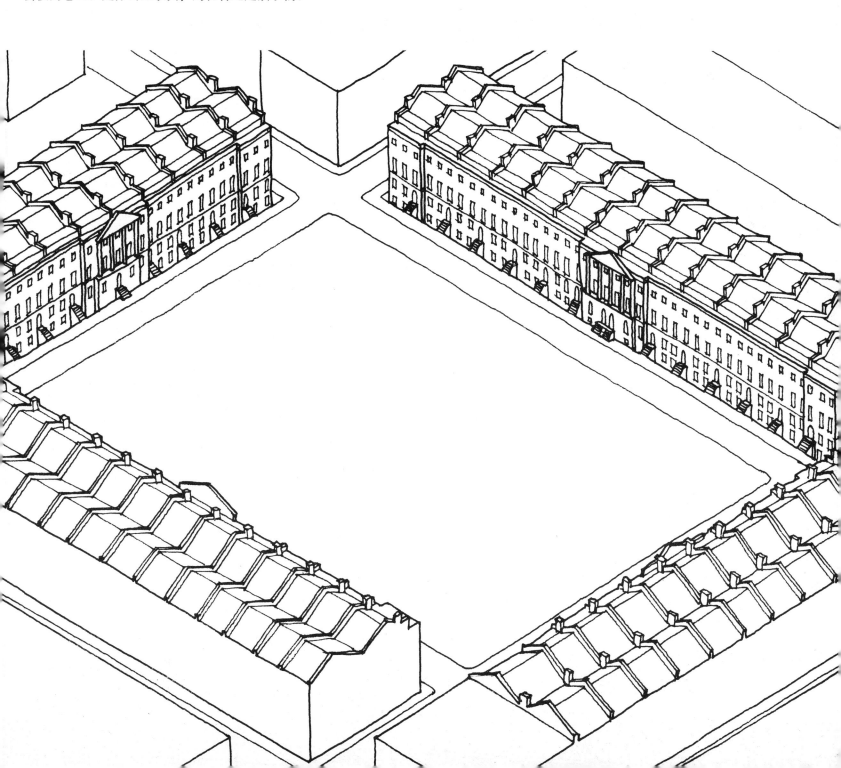

Running through the centre of **medieval Girona** in northern Spain is the Onya River, which is bound by houses and apartments that are typical of a Mediterranean city. The facades are painted in rich earth colours according to a palette created by Enric Ansesa, James J. Faixó, and the architects Josep Fuses and Joan Maria Viader.

乌纳河流经西班牙北部中世纪赫罗纳中心，这条河道两岸布满了典型地中海城市的住房和公寓。这些建筑立面的丰富色彩由恩里克·安塞莎，詹姆斯·J. 费克斯、建筑师何塞普·费塞斯和玛丽亚·比亚德尔创作的调色盘绘制而成。

为空缺的建筑设计新的立面,并且为这些所有建筑设计你自己的色彩。

These four examples of **gateways** to a city are largely symbolic as they constitute not only a landmark in the city, but also often relate to a particular historic event invoking ideals such as peace, freedom, liberty, protection and commemoration.

这里的四个城市门户案例都具有一定的象征意义，因为他们不仅构成了城市的标志，也常常与特定历史事件引起的相关理念联系在一起，比如：和平、自由、解放、防卫和纪念。

大拱门，圣路易斯，美国。埃罗·沙里宁，1965年。

勃兰登堡门，柏林。卡尔·圣歌达·朗汉斯，1971年。

印度门，德里。埃德温·路特恩斯，1931年。

平安神宫，牌坊，京都，日本。伊东忠太，1895年。

设计你自己的城市门户。这个门户或许和特别的历史事件有关,也可能由现代的形式和材料构成。

'What is the city but the people?' asked William Shakespeare, implying that the character of a place is formed by its citizens. These silhouettes of city dwellers represent some of the typical sorts of people who use public spaces and facilities. Along the foot of the page are icons that denote some of their daily requirements.

威廉·莎士比亚的名言"城市即人？"暗示出市民代表了一个地区的特征。这些城市居民的剪影代表了一些经常使用公共空间和设施的典型人群。页面底部的图标表示他们的日常需求。

城市居民：滑板爱好者、老人、年轻家庭、街头艺人

图标（上排）：自然、硬地面、雨棚、洗涤设施、饮用水、卫生间、桌子、休闲设施、睡眠空间
（下排）：人口稠密空间、人口稀疏空间、嘈杂空间、安静空间、骑行。

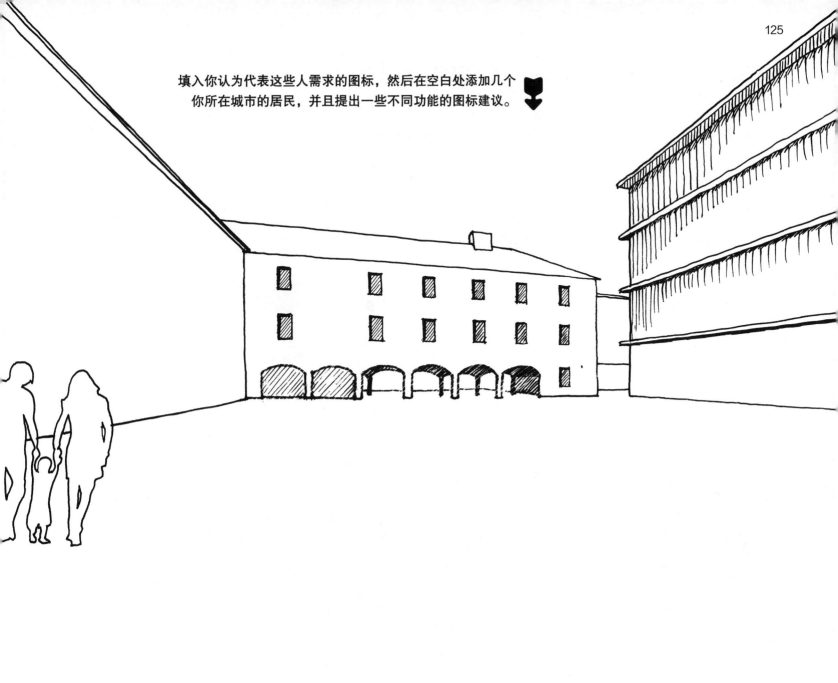

填入你认为代表这些人需求的图标，然后在空白处添加几个你所在城市的居民，并且提出一些不同功能的图标建议。

Many cities have a permanent theme park and funfair. **Coney Island** in Brooklyn developed as an amusement area in the late 1800s, and soon became one of the most popular theme parks in the USA. One of its main attractions was the 'Elephantine Colossus', built as a 34-room hotel in 1885. At 46 metres (150 feet) tall, the Colossus was so large that at one time it was the first thing that visitors would see upon arriving in New York. Unfortunately it was destroyed by fire in 1896 and only photographs and drawings now exist of this amazing structure.

许多城市都有一个永久性的主题公园和游乐场。布鲁克林的科尼岛在1800年代后期发展成为娱乐区域，而后很快成为美国最受欢迎的主题公园之一。它的主要的景点之一是一处有34个房间的"巨象"宾馆。因其高达46米且体量巨大，这个巨象曾一度成为游览者抵达纽约之后的首要观赏目标。但不幸的是它毁于1896年的一场大火，目前这个惊人结构仅有照片和图像被保存下来。

基于这里展示的平面和剖面轮廓展开想象，重新设计它的内部空间。哪里是入口和楼梯？有多少房间以及有些什么类型的房间？

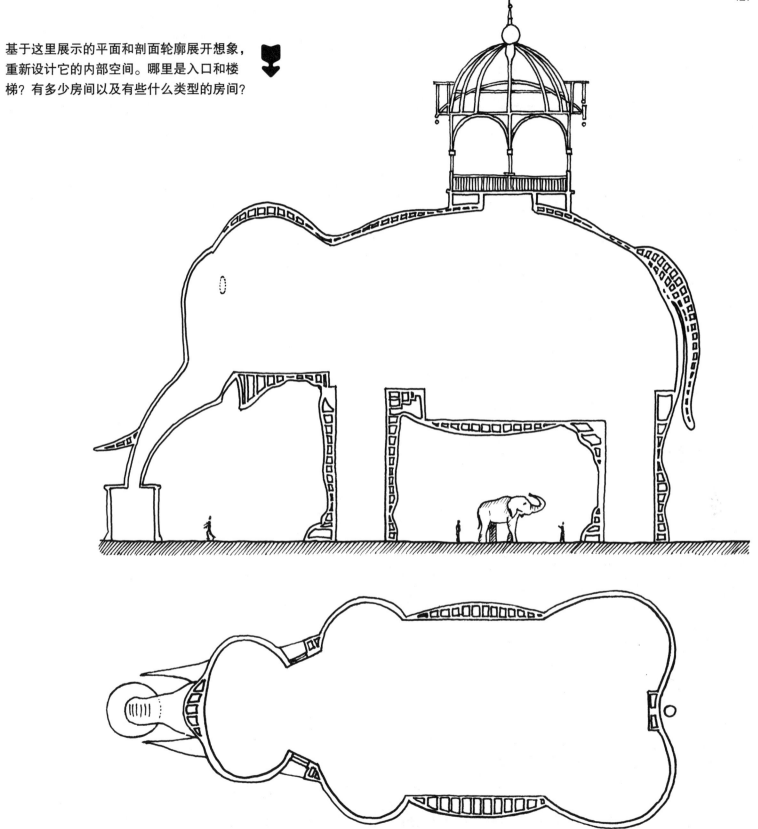

In 1922, famous Swiss-born architect Le Corbusier proposed a **'Contemporary City'** for three million inhabitants in response to the growing problem of the Paris slums. His 'Plan Voisin' (1925) envisaged the removal of two square miles of downtown Paris to provide a new housing quarter set within a large parkland.

1922年，著名的瑞士裔建筑师勒·柯布西耶提出了300万居民的"现代城市"规划，以应对巴黎日益严重的贫民窟问题。他在1925年的"伏瓦生规划"中设想清空巴黎市中心的2平方英里用地，在巨大的公园中建设新新住宅区。

你对这个清空的巴黎地区有什么建议？

勾勒出你的展示模型以及规划平面建议，包括沿塞纳河西岸延伸的商业街区、部分住宅区、文化和行政办公区。

These two buildings are examples of **museums** that have been dedicated to a single artist. Dalí said of The Salvador Dalí Theatre-Museum that 'I want my museum to be a single block, a labyrinth, a great surrealist object ... the people who come to see it will leave with the sensation of having had a theatrical dream'. The other example is dedicated to the Swiss painter Paul Klee and, while abstract in form, has a connection with Klee's famous quote that 'drawing is taking a line for a walk'.

这里有两个纪念一个艺术家的博物馆建筑案例。萨尔瓦多达利博物馆也是一座戏剧院，达利对这座博物馆的评价是，"我希望我的博物馆可以成为一个单独的街区，成为一个错综复杂的超现实主义物体……前来参观的人们将带着一种戏剧般的梦想离开"。另一案例反映了瑞士画家保罗·克利的风格，尽管采用了抽象形式，它却表达了其名言"绘画就是用一根线条去散步"的思想。

保罗·克利中心，伯尔尼，瑞士。
伦佐·皮亚诺，2005年。

萨尔瓦多达利剧院—博物馆，费格拉斯，西班牙。
乔奎姆·德·罗斯和亚历山大·博纳塔斯，1974年。

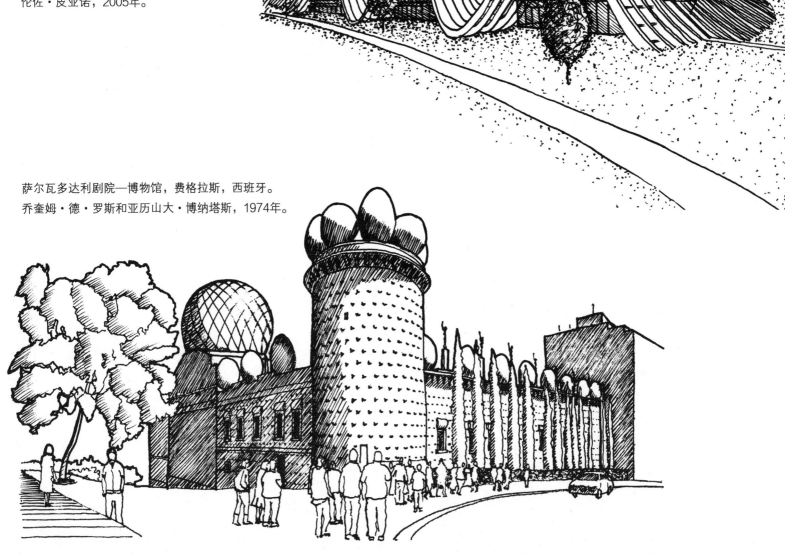

选择你喜欢的艺术家，并画出他们的博物馆可能的样子。考虑颜色、形式和材质，以及该艺术家的思想。

Many city skylines are characterized by the **roofs and towers** of their important religious and civic buildings. The Cathedral of Vasily the Blessed, better known as Saint Basil's Cathedral, in Moscow's Red Square has become both a focal point and a symbol of Russia's rich heritage.

许多城市的天际线因其重要的宗教或民用建筑的屋顶和塔楼而形成特色。位于莫斯科红场的瓦西里大教堂（又名圣巴西利亚大教堂）不仅形成了中心焦点，同时也成为了俄罗斯丰富遗产的象征。

瓦西里大教堂，莫斯科，俄罗斯。巴尔马和夫列夫，1561年。

切斯特菲尔德郊区教堂（扭曲的尖顶），德比郡，英格兰，14世纪。

勾勒出你自己的螺旋塔、尖塔或民用塔。
它是否和你居住的地方有什么文化关联？

哥本哈根证券交易所，丹麦，洛伦兹
和汉斯·范·斯提温克尔二世，1640年。

Shown here are a variety of typical **street profiles.** A street profile shows how the surrounding buildings interact and shape the character of the street. In my examples, you can see how the stepped facades, balconies, arcades, roof lines and basements all affect the streets' distinctive qualities.

这里展示的是几种典型的街道剖面。街道剖面显示出周边建筑互动并塑造街道特征的方式。在我的这些案例中，你可以看到阶梯立面、阳台、拱廊、屋脊和地下室是如何影响街道独特性的。

在提供的空间内绘制出你自己的街道剖面。

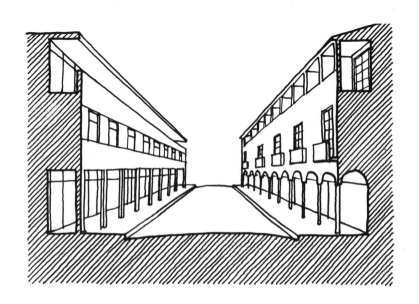

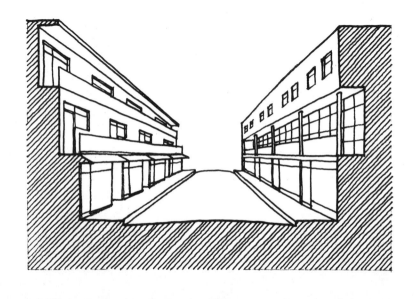

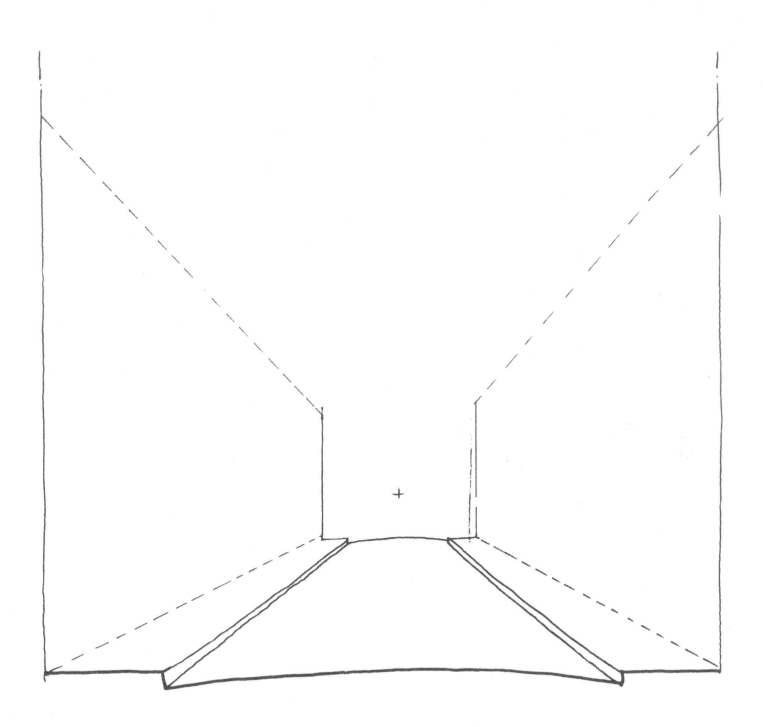

After World War II **new housing developments** were desperately needed in many European cities to replace the damage caused by the conflict. In post-war Britain, architects such as Peter and Alison Smithson and Ernö Goldfinger dedicated much of their practice to social housing and schools projects. Their buildings were uncompromising and were given the title of the 'New Brutalism', due to their raw appearance and use of unadorned concrete.

第二次世界大战之后，新的住房发展成为欧洲众多城市的迫切需要，人们需要它替代战争冲突造成的破坏。战后的英国，诸如皮特和戈德芬格等建筑师将他们的大部分实践用于社会住房和学校项目。这些建筑由于风格强硬，外观原始，并使用未装饰的混凝土而被赋予"新粗野主义"的称号。

在艾利克大厦轮廓内为社会住宅街区勾勒出你的创意草图。你会加入绿地空间吗？可持续技术呢？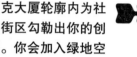

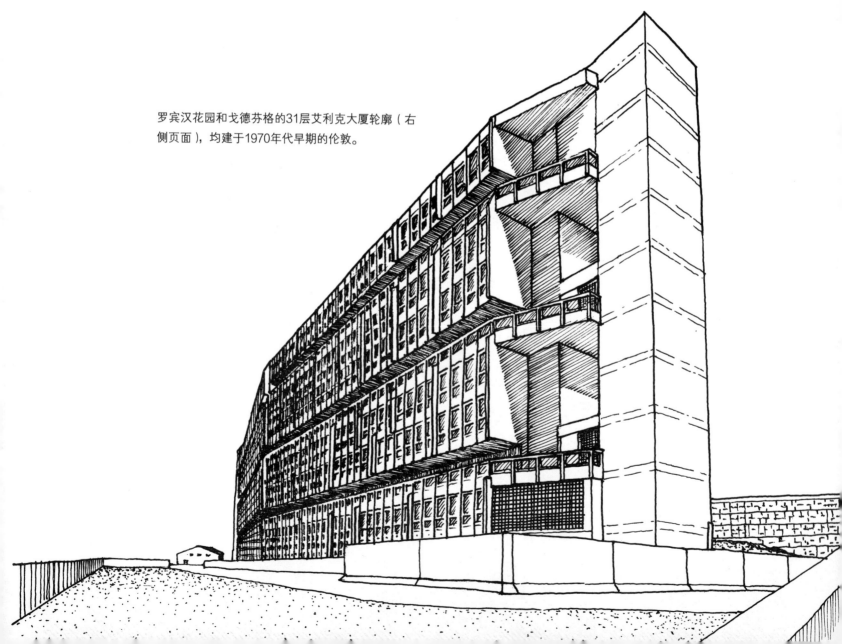

罗宾汉花园和戈德芬格的31层艾利克大厦轮廓（右侧页面），均建于1970年代早期的伦敦。

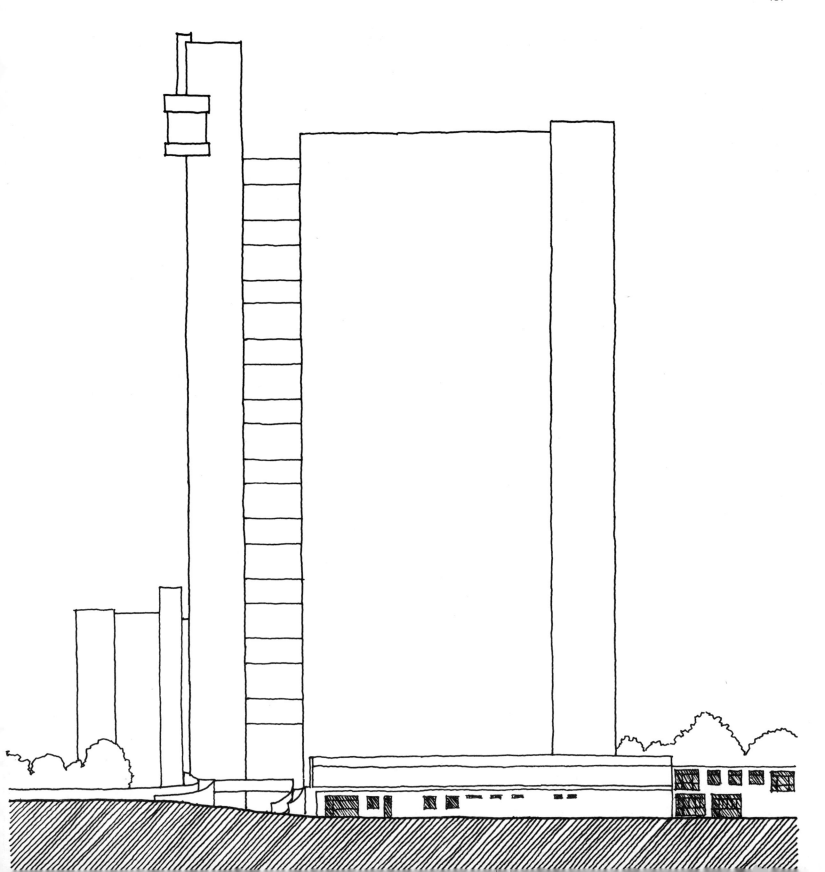

The designs of these modern **parliament buildings** utilized platonic forms (circles, squares, triangles) in the organization of their plans. In both buildings the circle is used as the 'democratic' meeting space for politicians to debate the future policies of their nations, with a variety of administration spaces surrounding the main hall.

这些现代国会建筑的设计在其平面组织中使用了圆形、正方形，三角形等理想形式。在每一个建筑中，圆形都被用作政治家讨论国家未来政策的"民主"会议空间，并以各种行政空间环绕主厅。

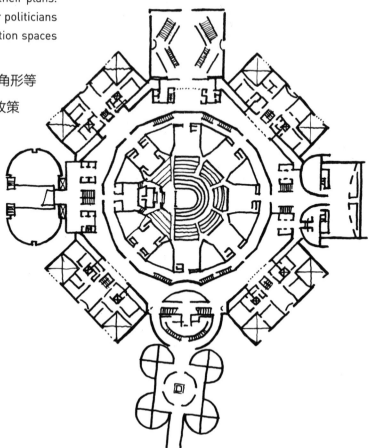

国家议会大厦、达卡，孟加拉国。路易斯·卡恩，1982年。

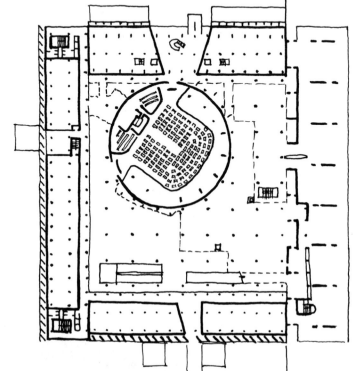

议会大厦，昌迪加尔，印度，勒·柯布西耶，1955年。

如果你将设计一个新的国会建筑，在使用下列形状的前提下，你将如何组织这些空间关系？主要会议空间是什么样子的，你将如何促进民主？

Famous for its sixteenth-century Renaissance and Baroque **merchant houses** is this square in Telc, Czech Republic. You will notice that all the dwellings share certain features – high gables, similar heights, arcades along the base and relatively symmetrical window patterns – while still being individual and unique.

这个位于捷克特尔奇的广场因16世纪文艺复兴和巴洛克式商业建筑而闻名于世。你会注意到所有的住宅都共享一定的特征，例如高山墙、相似高度、基座上的拱廊以及相当对称的开窗形式，但同时它们又具有各自个性和特点。

在给定的空间内设计并绘制两个立面草图，补足现有建筑，尝试保留原有素材中的一些共性特征。

'Boulevard of the History of Architecture'

is a drawing by Hans Dieter Schaal from the 1970s, in which he constructed a no-scale plan image of a street, juxtaposing some of the world's most significant buildings in a linear route.

"建筑史大道"是汉斯·迪特·史奈尔1970年代的作品,他在该作品中构建了一个无比例的街道平面影像,将世界上的一些最重要的建筑并置于一条线性路径上。

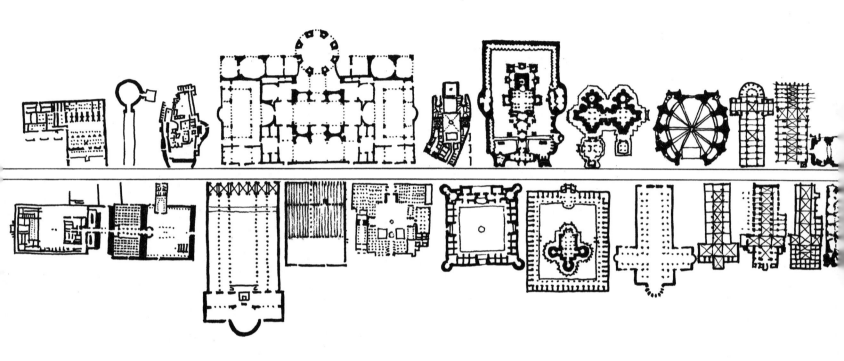

我增加了一些自己最喜欢的平面。你何不尝试着做相同的事情，在下图中从头开始创造属于你自己的大道。

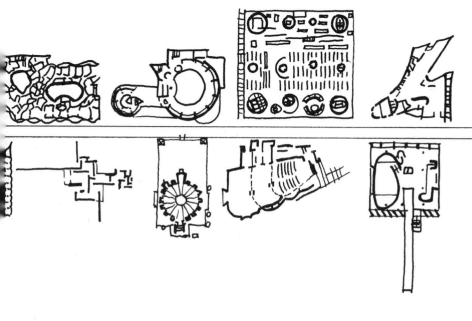

In their iconic 1972 book **'Learning from Las Vegas',** architects Denise Scott Brown, Robert Venturi and Steven Izenour analyzed the landscape of the Las Vegas strip. As it was a place and culture dominated by cars, they realized that signage was a very important factor for the casinos, motels and diners who were advertising themselves to their speeding customers. You will notice in the examples shown that the signage is composed of a much larger image or symbol that can be seen at long distance, and underneath there is more detailed information about the establishment and its services.

建筑师丹尼斯·斯科特·布朗、罗伯特·文图里和史蒂文·伊森尼在他们标志性的1972年著作《向拉斯维加斯学习》中，对拉斯维加斯大道的景观做了分析。作为一个汽车主导的地域和文化，他们意识到招牌对于赌场、汽车旅馆和餐厅向高速行驶的顾客宣传自己的重要作用。在这些展示的例子中，你能发现招牌由较大的图像或信号组成，以便于远距离识别，招牌下方则有关于公司和服务的详细信息。

为一家汽车旅馆或餐馆设计你自己的高速公路招牌。它能提供什么设施？招牌上都有什么信息？

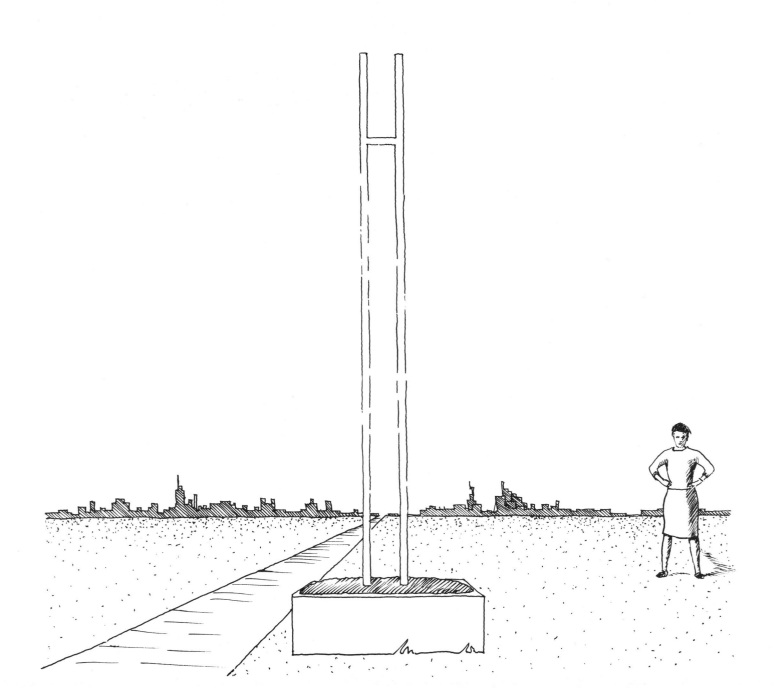

These two projects use **stacking blocks** on top of each other to create interesting spatial relationships. Designed for very different functions in different political environments, their construction is the one thing that they share in common.

这两个项目使用体块叠加的方式创造出有趣的空间关系。在不同的政治环境下它们可以有完全不同的功能，共同点则是具有相同的建造方式。

公路建设部（现在是格鲁吉亚银行），第比利斯，格鲁吉亚。19建筑师工作室和公路建设部长格奥尔基·察科娃，1975年。

栖息地67号住宅开发区，蒙特利尔，加拿大。摩西·萨夫迪，1967年。

使用相似的块状形式，为一个依赖简单堆叠形式的结构勾勒出一些草图想法。它的主导功能是什么？

The **roof terrace** of this unremarkable tower block in Beijing, China, became famous recently because the owner of the penthouse decided to convert his apartment's roof into a mountaintop retreat using rocks, rubble and shrubs. It took close to six years to complete but, as it did not have planning permission and was deemed to be unsafe, the owner was forced to remove it.

中国北京这个不起眼的塔楼屋顶露台最近变得出名起来，因为阁楼的主人使用岩石、碎砖和灌木将他的公寓屋顶变成了一座山顶别墅。这项工程花费了将近六年才得以完成，但是由于没有规划许可和安全性较差，业主被迫将其拆除。

如果你有一个塔楼顶端的阁楼，并且获准建设你期望的屋顶平台，你会如何设计？

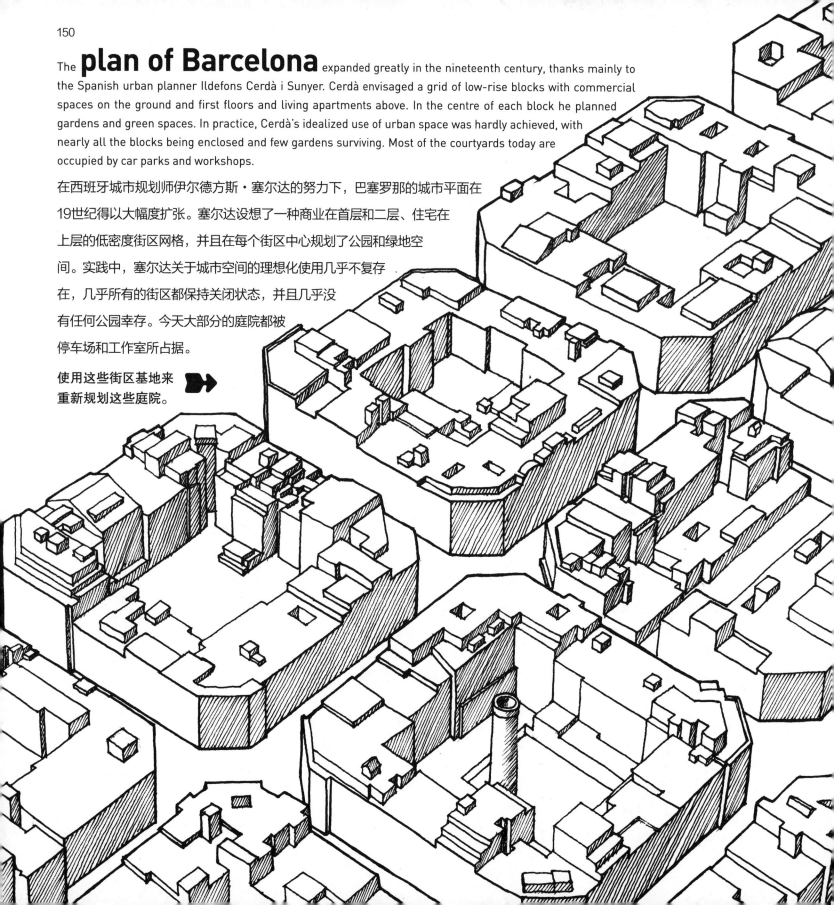

The **plan of Barcelona** expanded greatly in the nineteenth century, thanks mainly to the Spanish urban planner Ildefons Cerdà i Sunyer. Cerdà envisaged a grid of low-rise blocks with commercial spaces on the ground and first floors and living apartments above. In the centre of each block he planned gardens and green spaces. In practice, Cerdà's idealized use of urban space was hardly achieved, with nearly all the blocks being enclosed and few gardens surviving. Most of the courtyards today are occupied by car parks and workshops.

在西班牙城市规划师伊尔德方斯·塞尔达的努力下，巴塞罗那的城市平面在19世纪得以大幅度扩张。塞尔达设想了一种商业在首层和二层、住宅在上层的低密度街区网格，并且在每个街区中心规划了公园和绿地空间。实践中，塞尔达关于城市空间的理想化使用几乎不复存在，几乎所有的街区都保持关闭状态，并且几乎没有任何公园幸存。今天大部分的庭院都被停车场和工作室所占据。

使用这些街区基地来重新规划这些庭院。

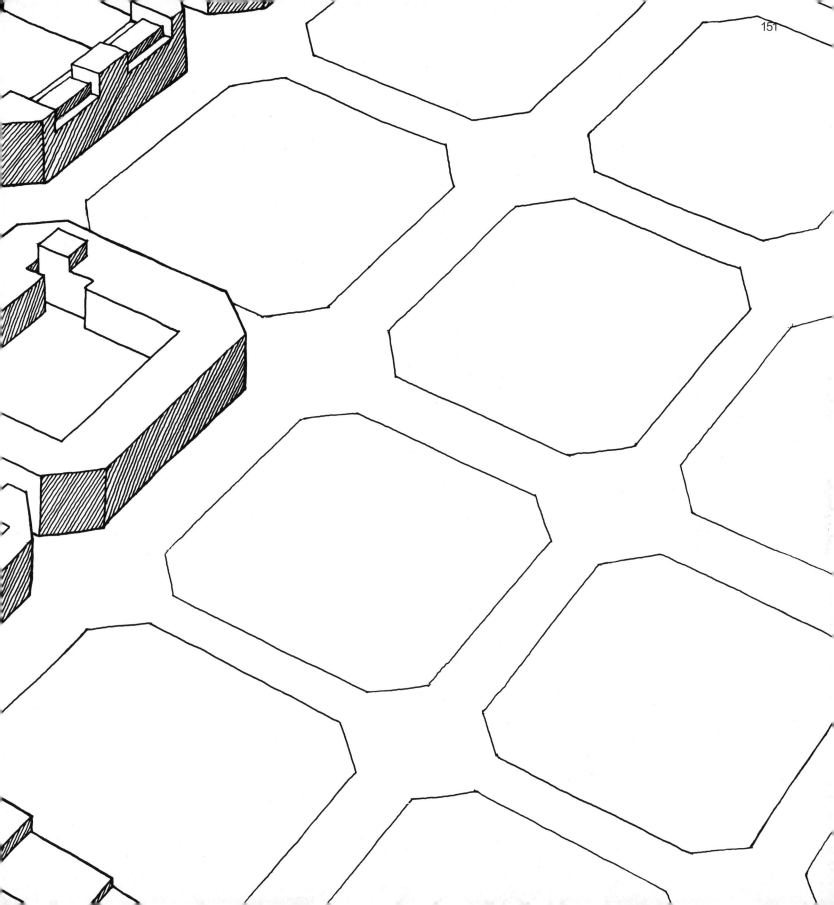

This **panorama collage** is composed of buildings from London and New York.

这是一幅由伦敦和纽约的建筑组成的全景拼贴画。

在这些图示中的著名大厦中加入你自己的标志性高楼。

In their influential book **'Collage City'**, published in 1978, Colin Rowe and Fred Koetter rejected the grand utopian visions of the past in favour of cities that embraced collision, superimposition, contamination and rich historical layering. Here I have used fragments of London, Paris, New York and Barcelona to form a new collage city. If they became one city, would it be called Ny-lon, Parcelona or maybe even New Barceldon? Would you include parks, avenues, squares and monuments?

柯林·罗和弗里德·科特在他们1978年出版的权威著作《拼贴城市》中，对过去支持城市接受冲突、重叠、玷污和历史分层的宏大乌托邦图景进行了反驳。在这里，我使用伦敦、巴黎、纽约和巴塞罗那的碎片组成一个新的拼贴城市。如果他们真的成为一个城市，它会被称作纽约−伦敦，巴黎−巴塞罗那，或者纽约−巴塞罗那−伦敦吗？你会加入公园、大街、广场和纪念碑吗？

绘制这些并与现有碎片合并，完成这个城市设计。 ➡

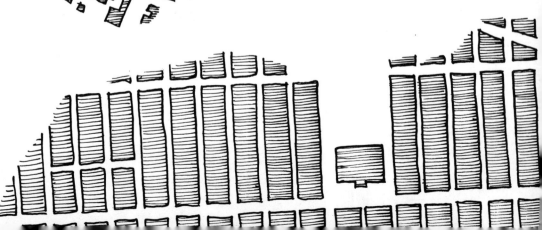

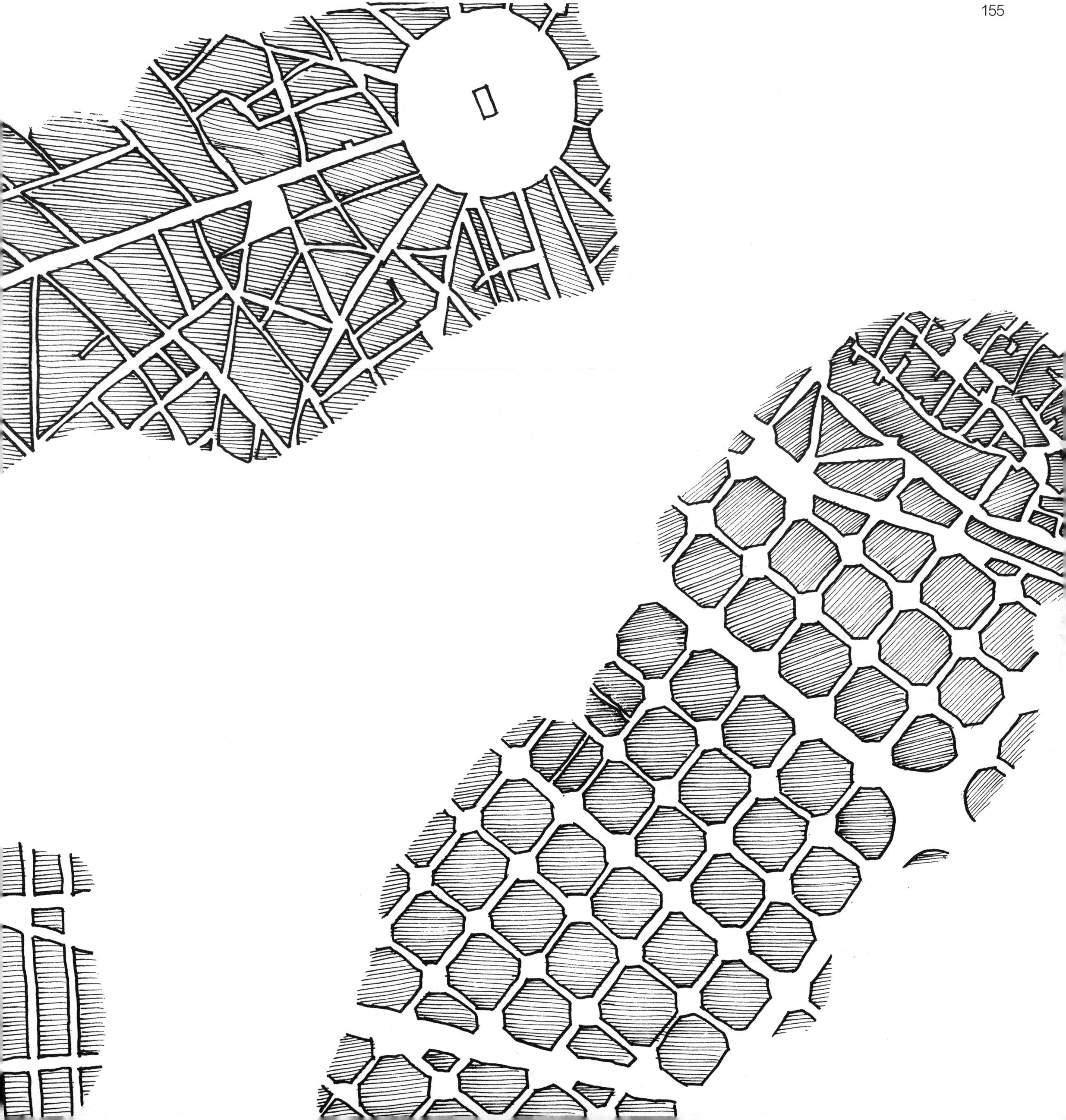

When the American architect **Louis Kahn** said 'The Street is a room by agreement', he was suggesting that streets were communal spaces that should serve the needs of the many. In recent times, proposals have been made for city centres to ban the use of private cars, thus opening up new possibilities for safer, cleaner and more active streets.

当美国建筑师路易斯·康说到"街道是协定的房间"时,他在暗示街道应该成为满足多数人需求的交流空间。最近,城市中心提出了许多禁止使用私家车的建议,从而为更安全、更整洁和更活跃的街道营造开辟了新的可能。

勾勒出一个有趣街道场景的个人创意,使用自行车道、阳台、咖啡馆和树等元素。

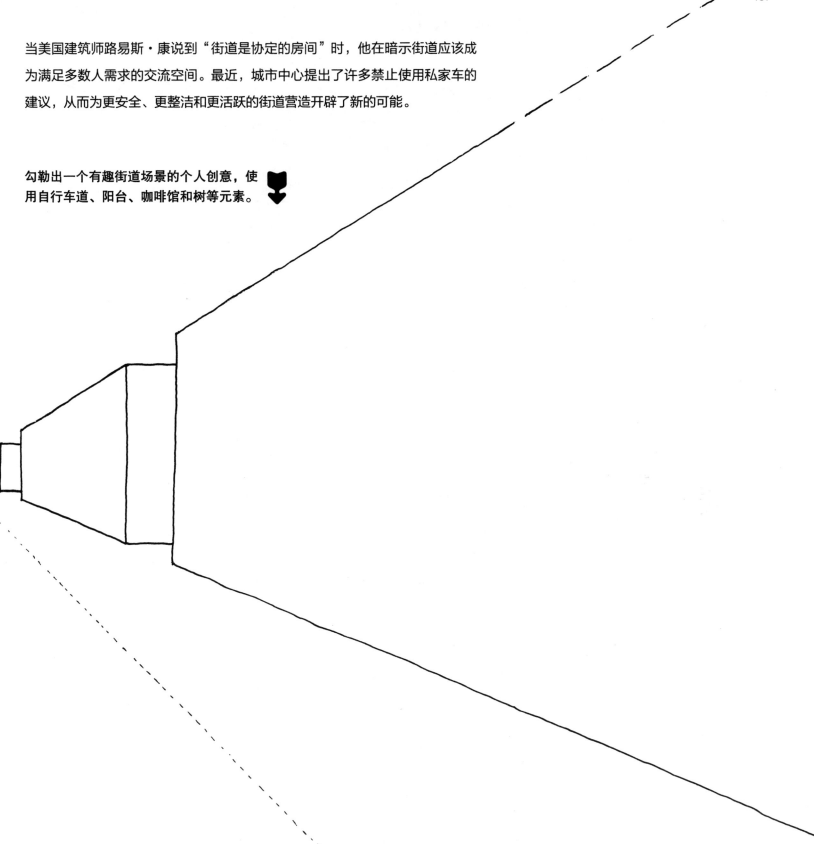

Located within Manhattan's East Village is **Alphabet City,** so called because its avenues A, B, C and D are the only streets in New York to have single-letter names. In the fictional cityscape below I made a name from the buildings.

曼哈顿东村有一个字母城,之所以叫这个名字是因为它们的A、B、C、D大街是纽约唯一拥有单一字母名字的街区。在下图这个虚构的城市景观中,我使用建筑制作了一个名字。

使用这些街区拼出你自己的名字,从而完成这个城市。

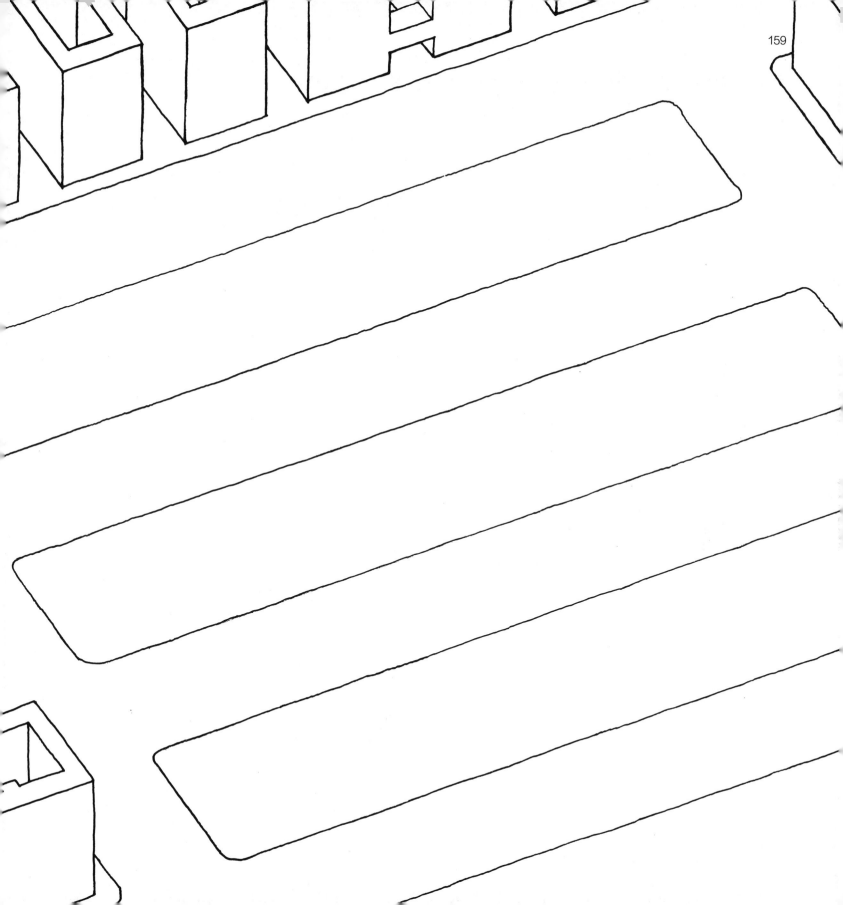

Credits & acknowledgements（致谢）

本书中所有的插图都是作者为此次出版特别创作的；其中一些基于建筑师和艺术家的原始画作，并且已经在合适的临近文本中予以注明。我们已经尽一切努力在所有的案例中获得版权拥有者的授权，但是如果有任何遗漏或错误，出版社将在此书后续版本中插入适当的感谢。

勒·柯布西耶的伏瓦生规划，巴黎，以及议会大厦规划，昌迪加尔©FLC/ADAGP和DACS，伦敦，2016年。

作者在此要对简·坦卡德（Jane Tankard）给予的所有支持、鼓励和耗时的反馈表示由衷的感谢。还要感谢菲利普·库珀、莉斯·法伯尔和Laurence King出版社Gaynor Sermon的帮助、支持和编辑工作，以及Newman and Eastwood的马特·考克斯对本书的设计。